AF801304

Nature, Nurture, and Neurodevelopment

Roya Kelishadi
Editor

Nature, Nurture, and Neurodevelopment

Comprehensive Overview

Springer

Editor
Roya Kelishadi
Department of Pediatrics
Arteriosklerose-Präventions-Institut
München, Germany

Child Growth and Development Research Center
Research Institute for Primordial Prevention
of Non-Communicable Disease
Isfahan University of Medical Sciences
Isfahan, Iran

ISBN 978-981-97-7432-6 ISBN 978-981-97-7433-3 (eBook)
https://doi.org/10.1007/978-981-97-7433-3

© The Editor(s) (if applicable) and The Author(s), under exclusive license to Springer Nature Singapore Pte Ltd. 2024
This work is subject to copyright. All rights are solely and exclusively licensed by the Publisher, whether the whole or part of the material is concerned, specifically the rights of translation, reprinting, reuse of illustrations, recitation, broadcasting, reproduction on microfilms or in any other physical way, and transmission or information storage and retrieval, electronic adaptation, computer software, or by similar or dissimilar methodology now known or hereafter developed.
The use of general descriptive names, registered names, trademarks, service marks, etc. in this publication does not imply, even in the absence of a specific statement, that such names are exempt from the relevant protective laws and regulations and therefore free for general use.
The publisher, the authors and the editors are safe to assume that the advice and information in this book are believed to be true and accurate at the date of publication. Neither the publisher nor the authors or the editors give a warranty, expressed or implied, with respect to the material contained herein or for any errors or omissions that may have been made. The publisher remains neutral with regard to jurisdictional claims in published maps and institutional affiliations.

This Springer imprint is published by the registered company Springer Nature Singapore Pte Ltd.
The registered company address is: 152 Beach Road, #21-01/04 Gateway East, Singapore 189721, Singapore

If disposing of this product, please recycle the paper.

Contents

Establishment of Neurodevelopment in the First 1000 Days of Life

Sadegh Mazaheri-Tehrani,
Maryam Bemanalizadeh, Parnian Poursafa,
and Roya Kelishadi

Abstract

The first 1000 days of life are the most imperative period for establishing the foundations of the nervous system in human beings. During this time, rapid proliferation, differentiation, and synaptogenesis of neurons occur, and different capabilities, including cognition, language, motor, and socioemotional, develop. Numerous genetic and environmental factors during pregnancy and early childhood might have a positive or negative impact on the neurodevelopmental process. According to the great impact of neurodevelopmental impairments on human quality of life and lifelong well-being, it is essential to take into account both harmful and beneficial factors in neurodevelopment. In this chapter, we aim to summarize these factors affecting the establishment of neurodevelopment and then discuss the importance of interdisciplinary approaches, besides the usual traditional ones (single-disciplinary), to discover the gaps in improving the neurodevelopmental process.

Keywords

Neurodevelopment · Brain building · Child development · First 1000 days of life · Gene–environment interaction

S. Mazaheri-Tehrani
Child Growth and Development Research Center, Research Institute for Primordial Prevention of Non-Communicable Disease, Isfahan University of Medical Sciences, Isfahan, Iran

Student Research Committee, Isfahan University of Medical Sciences, Isfahan, Iran

M. Bemanalizadeh · R. Kelishadi (✉)
Child Growth and Development Research Center, Research Institute for Primordial Prevention of Non-Communicable Disease, Isfahan University of Medical Sciences, Isfahan, Iran
e-mail: kelishadi@med.mui.ac.ir

P. Poursafa
Institute of Cell Biology and Neuroscience, Goethe University, Frankfurt am Main, Germany

1 The First 1000 Days of Life: The Golden Window of Brain Development.

The first 1000 days of life refers to the time from conception until the child's second birthday. It is the most important period of life because of its pivotal role in the establishment of the foundations of different aspects of health [1]. During this time, most of the systems, such as the immune system and brain, gain their appropriate functions. Thus, different factors during pregnancy and after birth may affect the child's lifelong health and well-being [2, 3]. This period is characterized by the rapid proliferation, differentiation, and synaptogenesis of neurons [1, 4].

© The Author(s), under exclusive license to Springer Nature Singapore Pte Ltd. 2024
R. Kelishadi (ed.), *Nature, Nurture, and Neurodevelopment*,
https://doi.org/10.1007/978-981-97-7433-3_1

The most rapid growth and development of the brain and nervous system occur during the last trimester of pregnancy and the first two years of life after birth. In the fifth month of pregnancy, the child's brain looks like a coffee bean, a smooth tissue with two lobes. By the third trimester, the child's brain develops into a complex structure with gyri and sulci, which looks like an adult's brain. At birth, the hippocampus and the visual and auditory cortices grow rapidly, and during the first year after birth, the language centers and the prefrontal cortex develop [1, 5, 6].

Studies have demonstrated different factors that adversely or positively affect neurodevelopment mechanisms during the first 1000 days of life [7–9]. Paying attention to these factors may decrease neurodevelopmental disorders (NDDs), which would affect the quality of life and lifelong well-being of each individual. Figure 1 demonstrates the effect of nature and nurture on the neurodevelopment process. According to a systematic review, the median prevalence of NDD is 7.6 per 1000 children in low- and middle-income countries, which is a considerable amount that imposes lots of cost on society and the child's family [10]. Here, we aim to discuss different aspects of this golden time in the neurodevelopment process.

2 Gene–Environment Interactions

Numerous genetic and environmental factors affect NDDs during the first 1000 days of life [11–13]. It is currently impossible to identify a precise etiological mechanism for NDDs. However, in the context of contemporary research, a shift from traditional nature–nurture approaches to a sophisticated framework of Gene x Environment (GxE) interactions has led to a more precise understanding of the complicated pathomechanisms. There is still limited understanding of the GxE interactions in NDDs, despite the fact that several studies are being conducted in this area. We first need to gain a broad understanding of the various complicated pathways of gene expression in order to highlight the genetic intricacy involved in NDD manifestations. At the gene level, DNA mutations, even a variation in a single nucleotide polymorphism, may result in a remarkable shift in the expressed 3-D molecular structure of proteins, potentially resulting in profound alterations in individuals' features. In some instances, the degree to which these mutations might impact children depends on the presence of compensating mechanisms in other implicated genes, which are called modifier

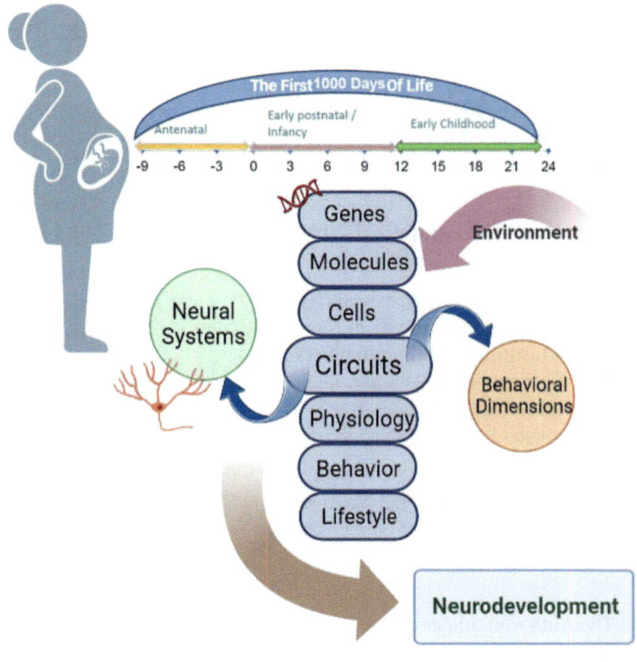

Fig. 1 Early life effects of nature and nurture on neurodevelopmental processes

genes, along the same biological route [14]. In addition, the interplay between genes depends on the temporal, spatial, and biochemical features of genes (i.e., the developmental stage, the location of interaction, and the active or resting state of proteins). In NDDs, considering the complicated features of genes, they often do not have a one-to-one relationship. Scientists have glanced beyond the genetics era in order to understand the precise mechanisms that underlie NDDs. Recently, scientists keep an eye out for the interplay of genes with risky environmental determinants. Understanding the idea of adaptability and developmental plasticity, in particular, demands consideration of the effects of genetic determinants, along with environmental hazards. Broadening knowledge of specific GxE mechanisms underpinning NDDs can lead to more productive risk-elimination strategies or preventive measures [15], but due to the complexity of the interactions, there is still much to be understood. On the one hand, research has shown that most neuropsychiatric disorders, such as NDDs, are moderately to highly genetically inherited [16]; however, scientists' endeavor to spot genes in charge of notable portions of variance in NDDs has been faced with several challenges and remains to be determined.

Based on the theory of differential susceptibility, individuals with genetic predispositions are more vulnerable to negative environmental determinants [17]. Furthermore, previous studies have highlighted the significance of a golden period—the first 1000 days of life—during which neural pathways, which stem from genetic data, might be potentially altered by environmental risk factors [18–20]. Similarly, the expression of some genes follows the same critical period, allowing the brain to rapidly develop and achieve its full functions during the first years of life [21].

3 Detrimental Factors for Neurodevelopment in the First 1000 Days

According to the defined term of the first 1000 days of life, various risk factors at different periods, i.e., prenatal, perinatal, and postnatal (first two years), may adversely affect the neurodevelopment process (Fig. 2).

3.1 Prenatal Factors

Maternal smoking during pregnancy is reported as an important detrimental factor for her offspring's brain development and intelligence quotient (IQ) [22]. Prenatal tobacco exposure (PTE) also causes lower brain volume and gyrification in adolescents [23]. However, there are controversies about the association between PTE and adverse neurodevelopmental outcomes among preterm infants [24, 25]. A longitudinal study on nonsmoker pregnant women indicated that exposure to environmental tobacco smoke (secondhand smoke) during pregnancy negatively affects psychomotor development within the first 1000 days [26]. Prenatal exposure to nicotine is reported to change brain structure by affecting neuron proliferation and differentiation. Besides, maternal smoking induces hypoxia and ischemia in the fetus, which will adversely affect the brain [27]. Research on electronic cigarette exposure during pregnancy has also found similar results [28].

Going along with PTE, studies have demonstrated that prenatal exposure to air pollution is associated with neurodevelopmental delays in children [29]. In utero exposure to air pollution is reported to have more harmful effects and cause more damage to the brain structure than such exposure during childhood [30]. Different air pollutants, including ambient particulate matter [31], nitrogen oxides [32], and carbon monoxide [33], are reported to be associated with impaired neurodevelopmental outcomes.

Maternal substance use during pregnancy has become more common in recent years. It is demonstrated that neurodevelopmental impairments might occur in infants with prenatal exposure to different substances such as opioids, cannabis, and alcohol [34]. Children of opioid-dependent mothers have poorer performance in various aspects of neurodevelopment, including cognition, psychomotor, language, and attention, compared to other children [35]. Alcohol consumption during pregnancy is associated with poor neuro-

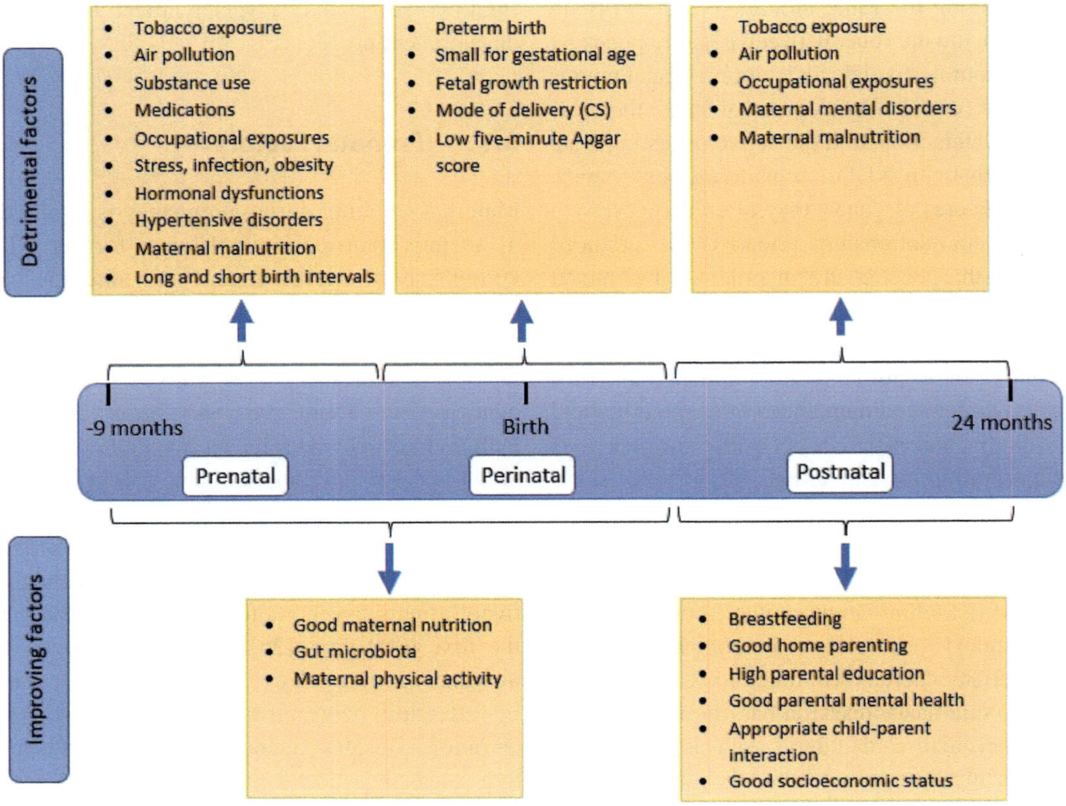

Fig. 2 Detrimental and improving factors in neurodevelopment during the first 1000 days of life

developmental and cognitive outcomes [36]. It is documented that there is no known safe amount of alcohol consumption during pregnancy [37], and a dose-dependent manner exists for psychological problems, somatic complaints, and attention disorders [38].

The consumption of some medications, including antidepressants, antipsychotics, and different analgesics during pregnancy, is reported to affect the child's neurodevelopment process [39]. It is imperative to be aware of the neurotoxic effects of medications during pregnancy and to replace them with safe ones.

Parental occupational exposure in the prenatal period can increase the risk of NDDs among their offspring. The brain of the fetus is extremely vulnerable to different toxins because the blood-brain barrier is not formed until the first few months after birth. In addition, the placenta cannot protect the fetus properly against various toxins. Prenatal exposure to solvents or pesticides might increase the risk of attention deficit hyperactivity disorder (ADHD). Moreover, prenatal pesticide exposure is associated with motor and cognition disorders [40]. Prenatal exposure to different metals, including aluminum, cadmium, molybdenum, lead, and mercury, is reported to be associated with negative neurodevelopmental outcomes in early childhood [41, 42].

There are controversies about the effect of mobile phone (MP) use during pregnancy on the neurodevelopmental outcomes of offspring. A population-based cohort in Spain suggested early adverse neurodevelopmental outcomes as a result of maternal MP use during pregnancy [43]. However, other studies found no significant association [44–46]. One study even indicated that maternal MP use during pregnancy lowered the risk of language and motor skills disorders [47].

Longitudinal studies are necessary to assess such potential adverse effects of mobile phone use.

Maternal stress during pregnancy is an important risk factor for offspring's NDDs, including ADHD, autism spectrum disorder (ASD), and schizophrenia. Maternal infection during pregnancy is also another detrimental factor for neurodevelopmental outcomes [48, 49]. Various mechanisms have been suggested for NDDs due to antenatal maternal stress or infection. Maternal immune activation (MIA), because of infections or autoimmune diseases in mothers, or an increase in inflammatory markers, especially interleukin-6 (IL-6) and IL-17, are the main underlying pathways [50–52]. Several epidemiological studies have demonstrated an increase in NDDs during pandemics or epidemics of infections [51]. Of note, studies have suggested a neurotropism effect for SARS-CoV-2, as the greatest twenty-first-century pandemic, but further research is needed to understand the underlying mechanisms for COVID-19-induced NDDs [53, 54].

Besides the mentioned risk factors, maternal obesity increases the chance of NDDs in infants by inducing systemic inflammation in the mother's body, negatively affecting the development of the dopaminergic system, and inducing epigenetic alterations [55]. Maternal obesity [56] and a high-fat diet (typically a Western diet) [57] are reported to be associated with an increased risk of NDDs such as ASD, ADHD, cognitive impairments, and cerebral palsy in offspring. Moreover, the results of a meta-analysis revealed that prepregnancy obesity also increased the risk of ADHD, ASD, and emotional and behavioral problems [58].

Clinical and subclinical hypothyroidism (CH, SCH) are among the most frequent hormonal dysfunctions during pregnancy. Studies have demonstrated that both CH and SCH are associated with an increased risk of intellectual and cognitive disorders in offspring during the first two years of life [59]. However, the effect of maternal CH and SCH on ADHD and ASD is still unclear. Moreover, there is no evidence that levothyroxine treatment during pregnancy is useful for lowering the risk of intellectual disability in offspring [60]. Another hormonal dysfunction that may affect the neurodevelopment of infants is gestational diabetes mellitus. Infants of diabetic mothers showed lower scores for mental and psychomotor development indexes [61]. Of note, there is no difference in 2-year-old infants' neurodevelopmental outcomes in mothers treated either with metformin or with insulin [62]. Neonatal hypoglycemia is a common metabolic complication in newborns of diabetic mothers [63], and in turn, it is associated with an increased risk of neurodevelopmental impairments in mid-childhood, i.e., children aged 6–11 years [64]. However, there are doubts about the effect of neonatal hypoglycemia on early childhood (2–5 years old) neurodevelopment [64, 65].

Hypertensive disorders of pregnancy are a well-established risk factor for adverse neurodevelopmental outcomes and different NDDs during the first 1000 days of life [66–68], though there are doubts about their long-term neurodevelopmental outcomes [69].

According to the World Health Organization (WHO), it is recommended to interval at least two years after a live birth before attempting to become pregnant again [70]. Both long (more than five years) and short (less than one year) intervals between pregnancies are reported to be associated with an increased incidence of neurodevelopmental disorders, such as ASD, in the offspring [71].

According to a systematic review, in developing countries, singletons showed better neurodevelopmental outcomes than twins, whereas such differences are not documented in developed countries. Such differences might be because of better health care in high-income countries. Overall, studies that assessed early neurodevelopmental outcomes reported no significant difference between singletons and twins [72]. Results of another systematic review indicated that monochorionic twins have poorer neurodevelopmental outcomes than dichorionic twins. Moreover, monochorionic infants with twin-twin transfusion syndrome reported more neurological impairments. Fetoscopic laser surgery lowered the risk of poor neurodevelopmental outcomes in monochorionic twins with this syndrome [73].

3.2 Perinatal Factors

According to the WHO, it is estimated that 15 million preterm births occur each year worldwide, and approximately one million babies die per year because of related complications. Preterm birth is categorized into four groups based on gestational age: extremely preterm (less than 28 weeks), very preterm (28–32 weeks), moderate preterm (32–34 weeks), and late preterm (34–36 weeks) [74]. All categories of preterm birth are important risk factors for either short-term or long-term neurodevelopmental impairments and disorders [75–78]. Because of the high prevalence of preterm birth worldwide on one hand, and the irreparable neurological consequences on the other hand, early identification of possible neurodevelopmental impairments in high-risk babies is imperative. Screening examination, besides magnetic resonance imaging of the brain, is a proposed method for the early identification of possible disorders. The best way to prevent these undesirable outcomes is to prevent the occurrence of preterm births. Developmental interventions early after the discharge of the preterm infant, parent–child interactions through playing, and increasing parents' knowledge about the early signs of impairments are the proposed interventions to improve neurodevelopmental outcomes [79].

Small for gestational age (SGA) is defined as a birth weight lower than the tenth percentile for gestational age. Based on fetal growth, SGA is classified into two groups: SGA infants with normal intrauterine growth and SGA infants with fetal growth restriction (FGR). Most of the SGA infants have FGR and vice versa. However, SGA is not an imperative factor for FGR, because some neonates with FGR have normal birth weights [80]. SGA is a well-established risk factor for different neurodevelopmental outcomes, including cognition and behavioral impairments [81]. Among SGA infants, those who are FGR are reported to have a higher risk for adverse neurodevelopmental outcomes [82, 83]. On the other hand, there are controversies about the association of large for gestational birth with neurodevelopmental outcomes [84–86].

The delivery mode is a factor that may affect neurodevelopmental outcomes. Cesarean section is reported to be associated with a decrease in gross and fine motor skills, as well as language development, compared to vaginal delivery [87]. Cesarean section also might increase the risk of NDDs such as ASD and ADHD, as well as eating disorders [88]. However, this negative effect of CS on neurodevelopmental outcomes may be influenced by different maternal and socioeconomic factors [89, 90].

The clinical condition of the newborn immediately after birth is reported to affect its neurodevelopmental outcomes. The Apgar score is used as a rapid tool for assessing health status based on respiration, heart rate, skin color, reflex irritability, and tone of muscles. The Apgar score is measured at the first minute and fifth minute after birth, and its scores range from zero to ten. A low five-minute Apgar score (mainly defined as scores below seven) is associated with an increased risk of mortality and different neurodevelopmental impairments. Studies have demonstrated that even five-minute Apgar scores of 7–9 can increase the risk of neurodevelopmental delay [91, 92].

3.3 Postnatal Factors

Some detrimental factors after birth are the same as prenatal factors. Exposure to secondhand smoke [93], ambient particulate matter [94], and environmental toxins such as polycyclic aromatic hydrocarbons and organophosphorus insecticides [95], are reported to have harmful impacts on early childhood neurodevelopment.

Maternal mental disorders such as depression, anxiety, and stress, especially during the postnatal period, are associated with various neurodevelopmental aspects of offspring, including language, motor, intellectual, and behavioral domains [96]. However, when evaluating mental disorders separately with each domain of neurodevelopment, the results become less evident and inconsistent [97, 98]. Different confounding factors, such as socioeconomic status, parental age and education, and home environment, may explain these controversies [96].

4 Factors Improving Neurodevelopment in the First 1000 Days

Some parameters may facilitate the neurodevelopment process (Fig. 2). Shedding attention to these improving factors may be effective in preventing various neurodevelopmental impairments.

4.1 Prenatal Factors

Maternal nutrition during pregnancy affects the neurodevelopmental process of the offspring. Maternal malnutrition, ranging from undernutrition to overnutrition during pregnancy, negatively affects different neurodevelopmental processes, which predisposed the fetus to different NDDs. Sufficient amounts of fatty acids, including polyunsaturated fatty acids, docosahexaenoic acid, and arachidonic acid, in the maternal diet are reported to have beneficial effects on the neurodevelopmental processes through the formation of neural cell membranes, myelination, and synapse maturation [99–101]. However, there is limited data on the effect of maternal supplementation on neurodevelopmental outcomes [102–104]. Proteins are another important component of the maternal diet, especially during the second and third trimesters for neurodevelopmental processes. A daily intake of 50 g of protein is recommended during pregnancy [99]. Different maternal micronutrients, such as folate, vitamin B12, vitamin D, iodine, and choline, are also reported to play a role in the neurodevelopmental process of the fetus [99, 101]. However, there is limited evidence about the exact effects of some nutrients during the prenatal period and their consequences in neurodevelopmental processes. Iron is an essential element with an established role in the neurodevelopmental process, but studies demonstrate inconsistent results regarding the effect of iron deficiency on neurodevelopmental outcomes [105, 106]. Surprisingly, iron supplementation during pregnancy is not associated with offspring neurodevelopmental outcomes [107].

The impact of maternal nutrition on the infant's gut microbiota is a newly investigated mechanism for the effect of maternal nutrition on neurodevelopment [108]. Gut microbiota affects various functions in the nervous system, including neurogenesis and myelination, which strengthens the effect of the gut-brain axis in neurodevelopment [109].

Maternal physical exercise during pregnancy has different benefits for both mother and offspring. Decreasing the risk of maternal gestational diabetes, hypertension, excess gestational weight gain, and perinatal depression are among the benefits for mothers [110, 111]. Improving neurodevelopmental outcomes, including intellectual and language skills, are the main benefits of maternal physical activity during pregnancy for her offspring [112, 113].

4.2 Postnatal Factors

Human milk provides optimal nutrients for infants during the first two years of life. According to the WHO statement, it is recommended to exclusively breastfeed for the first six months and then provide complementary foods together with breastfeeding until the second birthday or beyond [114]. Studies have demonstrated that breastfeeding has short-term and long-term neurodevelopmental benefits and is associated with better intellectual outcomes compared to formula feeding [115, 116]. Infants who were breastfed for more than six months are found to have better intelligence performance than other children [117]. In addition, human milk has slightly beneficial effects on the neurodevelopment of preterm infants [118]. However, there are doubts about the effect of the amount of human milk consumption on the neurodevelopmental outcomes of preterm babies [119].

Different family environment factors during early childhood are reported to shape the neurodevelopment of babies. Home parenting, high parental education, good parental mental health, and appropriate interactions between parents and children can promote neurodevelopmental outcomes in babies [120, 121]. Of note, low socioeconomic families are at higher risk of neurodevelopmental impairments in offspring. This effect may be due to several co-existent adversities such as nutritional deprivations, psychosocial stress, tobacco and toxic chemical exposure, as well as reduced breastfeeding, which are more frequent among low socioeconomic families [121]. However, there are doubts about the extent of such effects of parenting intervention programs on neurodevelopmental outcomes [122, 123].

5 Necessity of Interdisciplinary Research to Understand Brain Development in First 1000 Days of Life

Despite the growing number of scientific advances over the last century, many aspects of brain development and NDDs remain unknown. Since the nature of the human brain and its neural networks is intricate, traditional approaches (single-disciplinary approaches) are not enough in this field of science. To focus on the health-related issues of the new century, both single-disciplinary and interdisciplinary concepts should be considered. In this section, we will focus on some interdisciplinary contributions coming together to answer specific questions or solve particular problems regarding brain development and NDDs in early childhood. Integrated efforts among various disciplines can lead to investigate the fundamental mechanisms of brain function and to develop strategies for prevention, early detection, and intervention in NDDs. There is a vast majority of disciplines, including but not limited to pediatrics, neurology, psychiatry, psychology, cellular and molecular biology, radiology, neuroscience, artificial intelligence, bioengineering, and technology, all of which contribute to new findings in various aspects of brain development and NDDs. Several interdisciplinary programs have demonstrated the added value of interactive efforts. An interdisciplinary approach assists physicians in overcoming the remarkable problems faced in the current context for NDDs' prevention, assessment, and treatment, particularly in the first 1000 days of brain development [124, 125]. Diverse strategies are used in interdisciplinary research. This section consists of some examples of recent translational research on brain development and NDDs, integrating clinical findings with various disciplines, which can result in developing new perspectives.

The early occurrence of NDDs and the delay in diagnosis—which is determined based on symptom onset—pose a fundamental obstacle in spotting the cause and administering the appropriate treatment. Moreover, for almost all NDDs, therapies still depend on behavioral or educational programs that alleviate accompanying symptoms rather than targeting the main etiology [126]. Consequently, almost all individuals with NDDs suffer from poor outcomes in adulthood, which can seriously affect their quality of life [127]. Furthermore, the clinical heterogeneity of NDDs and the complicated nature of brain development make all steps of clinical practice a time-consuming and pricey process, hampering the implementation of personalized medicine. Given such challenges, recent advances in genetics and genomics may lead to new discoveries. First and foremost, genetic screening can provide prompt surveillance and early management, even prior to symptom onset. In addition, understanding the biological etiology of a disorder may shed light on the involvement of a particular biochemical mechanism that can result in targeted pharmacological therapies through repurposable or novel drugs. Moreover, future gene therapy studies can be conducted on single-gene genetic disorders. The recent approval of Zolgensma for spinal muscular atrophy has shown some promising results for future products in NDDs [128].

Thanks to remarkable developments in genomics, including whole-exome sequencing (WES) and whole-genome sequencing (WGS), scientists have been able to detect a vast number of mutations underlying NDDs. Considering the hundreds of risk genes that have been detected for NDDs as well as the wide spectrum of clinical manifestations, a precise diagnosis needs the genotype in addition to the phenotype of patients. Recent years have witnessed the rapid growth of accurate technological and methodological advancements in genetics and genomics, enabling prominent recognition of mutations that play a role in NDDs. Based on previous studies, NDDs are reported in 1–4% of children worldwide [129] and are caused by mutations at over one thousand loci [130]. Figuring out the causes of NDDs faces many obstacles, such as detecting the heritable genetic factors, identifying the individual risk factors for NDDs, and determining the precise underlying pathways [131]. Furthermore, since NDDs are accompanied by cognitive and behavioral impairments, genetic data can help in selecting the best behavioral interventions [132]. Nevertheless, in spite of the recent advances, guidelines advocating genetic testing for patients with NDDs are still not carried out in clinical practice. This is either because of the lack of facilities or the few numbers of trained clinical staff who are able to interpret the genetic findings. Therefore, a good dissemination of current genetic data regarding NDDs to medical staff and health care providers is needed. In addition, genetic counseling should advise parents regarding the assessment for possible recurrence in their next children [133]. On the other hand, the massive data management issues due to the abundance of data generated in genetic investigations and genomic analyses need more powerful computational approaches, which have been possible with systems biology and machine learning strategies. In addition, this issue raises ethical and legal concerns stemming from philosophy, law, and ethics. Therefore, not only is biology a trending discipline in NDDs, but it has also expanded to other disciplines [124].

Understanding the underlying mechanisms and advances in new treatments is not possible without preclinical studies. Therefore, generating models of NDDs that can partially mimic disorder features is definitely necessary for future research.

A perfect model for preclinical studies should have satisfactory construct validity, which means the model must simulate the genetic impairment that can result in the disorder. Face validity and predictive validity should also be considered; this means the model's presentations should mimic the human symptoms, and the response to a particular treatment should be similar in both the model and the patient. Several systems, such as cells, rodents, and primates, have been utilized to simulate NDDs to some extent in order to understand the underlying pathways contributing to NDD presentations. The mouse has been widely used to represent NDDs with established genetic etiologies [134–139]. Mice considerably share their genomic information with humans (95–98%) and reproduce at a relatively rapid rate. Moreover, they are more cost-effective in comparison with other animal models, and these items allow researchers to modify their genome. However, they also have significant drawbacks. For instance, it is difficult to test some higher brain functions, including language and facial recognition, in huge screens. For assessing these conditions, nonhuman primates are preferred to represent complicated behavior and higher brain functions [140], whereas zebrafish and invertebrates can be employed for high-throughput genetic screens [141]. In addition to animal models, the in vitro reprogramming of stem cells (i.e., human embryonic stem cells (hESCs) and human induced pluripotent stem cells (hiPSCs)) has made it possible to generate and investigate human neurons. Scientists have stimulated several neuronal synaptic impairments for monogenic forms of NDDs, such as Rett syndrome [142], Prader-Willi and Angelman syndromes [143], Phelan-McDermid syndrome [144], Fragile X syndrome [145], Down syndrome [146], and Timothy syndrome [147]. Some advantages of stem cell-based models include experimental tractability, the ability to mimic disorders directly from affected patients, and access to an infinite reservoir of cells. In contrast, the great variability across iPSC clones, the immaturity of neurons generated in vitro, the absence of complex connections and networks between neurons as well as other types of cells such as astrocytes, and the 2D modeling of cells are some of the most critical drawbacks of iPSC-derived neurons. Fortunately, scientists have recently developed new strategies for generating 3D cortical organoids (mini-brains/spheroids), providing potential to investigate more the brain development in vitro [148, 149].

Within the various disciplines, neuroscience is a vivid demonstration of an interdisciplinary approach that bridges the efforts of researchers from diverse sectors to address common scientific challenges and to obtain perspectives on crucial questions about nervous system development. It is a dynamic field that continues to incorporate additional fields, including informatics and molecular biology. The growth of neuroscience has been so tremendous and broad that it is beginning to generate various subdisciplines. One such subdiscipline is developmental neuroscience, which is, by itself, becoming a distinct field of study [124].

Neurodevelopmental engineering is another new interdisciplinary field of study at the center point of developmental neuroscience and bioengineering, essentially focused on developing novel technologies for recognizing neurobiological pathways of human brain function, behavioral simulation of neurodevelopmental milestones achieved by infants from birth onwards, and designing new mechatronic and robotic systems for implementation on children in minimally structured environmental conditions. This interdisciplinary field can be applied in new clinical guidelines and standards for early diagnosis, functional assessment of NDDs, and new generations of educational, interactive toys that can provide appropriate stimuli and guidance to support the physiological neurodevelopment process. This discipline can be helpful in the early diagnosis of NDDs and behavioral monitoring of infants by novel mechatronic equipment in different forms, such as wearable or embedded into toys. Neurotechnological devices can resolve many hurdles, such as delays in diagnosis and loss to follow-up in infants in general and preterm infants who are at higher risk of NDDs in particular [150–152].

Likewise, telehealth can help clinicians with better decision-making throughout the clinical practice, from diagnosis to treatment and rehabilitation. Telehealth has the potential to simplify early screening, decrease delays in diagnosis, increase treatment accessibility and service delivery efficiency, as well as assist in neurodevelopmental monitoring in a cost-effective setting. Recent findings revealed some promising results in the effect of telehealth programs on the assessment, management, and monitoring of different NDDs [153–156].

In recent years, the number of brain imaging studies on NDDs has exploded. Neuroimaging seeks potential in identifying neurological markers that can result in earlier and reliable diagnoses by identifying the involved neural networks in NDDs. Although these markers do not address the causes of the disorder, they can reliably predict the neural signature and accurately classify children with such disorders [157]. The necessity for decisive biomarkers derives from the limitations of NDDs diagnosis based on behavior. A more objective and quantifiable biomarker could reduce missed diagnoses, lead to earlier diagnosis, and identify high-risk children who would benefit from early preventive treatment. Another aim of neuroimaging is to identify abnormal brain processes underlying disorders that can be treated. In addition, neuroimaging can quantify the impact of interventions on brain circuits. This objective has translational impacts and can utilize the results from fundamental neurobiology research for therapeutic modalities. However, technical issues in neuroimaging for infants, such as using pediatric brain templates for normalization, head motion correction, delineating convoluted developmental circuits, and overcoming smaller sample sizes, should be considered in future neuroimaging studies [158]. Considering the big data revolution in neuroimaging studies, artificial intelligence and machine learning models can bring a whole new transition in our understanding of NDDs. These novel approaches to clinical diagnoses have the potential to enhance conventional methods that use behaviors and symptoms, with the hope of pushing forward personalized intervention plans. Therefore, the application of machine learning models to NDDs can potentially enhance screening programs and early detection, advance our knowledge about comorbidity complexity, and improve the detection of biomarkers for clinical research and drug discovery [159].

Most national and international guidelines recommend the inclusion of interdisciplinary assessment in the diagnostic workup for complex disorders such as NDDs. Hence, interdisciplinary assessment should always be part of the NDD diagnostic process in children [160]. Over recent decades, one of the fundamental efforts in NDD research has been devoted to biomarker discovery, helping clinicians make more accurate and efficient diagnoses according to the information about fundamental mechanisms rather than dependence on symptoms only [161, 162]. In general, a biomarker could be any index or test, ranging from genetic markers and biochemical assays to brain imaging scans, electrophysiological investigations, eye-tracking and behavioral tracking read-outs, or cognitive tests. Therefore, all of these methods, alone or in combination, have the potential to be recognized as biomarkers, first for research applications and then for clinical application in the foreseeable future [163, 164]. Moreover, novel therapies, from behavioral interventions to pharmacotherapies or even gene therapies, need integrated interdisciplinary collaborations to provide precision medicine for children with NDDs in the early years of their lives.

Extraordinary etiology variations and involvement at many levels ranging from genes to neural networks suggest the disparities in our current understanding in the field of NDDs and brain development. Considerable advancements in human health depend on the interdisciplinary research that needs close collaboration between biomedical, behavioral, and clinical scientists. Interdisciplinary research has a crucial role in the scientific enterprise. Although research in single disciplines has been established and will be progressed, disciplines should come together to answer questions that originate from multiple aspects in NDDs.

6 Conclusion

Numerous factors affect the neurodevelopment process, whether before or after birth, during the first 1000 days of life. According to the huge impact of brain development on human quality of life, it is imperative to make parents aware of the importance of this golden period and the factors that increase the chance of neurodevelopmental impairments or protect against them. Most of the detrimental factors, such as maternal tobacco exposure, obesity, and a high-fat diet, are preventable, while some genetic factors are unmodifiable. However, the exact effect of some factors, such as mobile phone use, being a twin, and being large for gestational age, are not yet well established. Of note, the effect of some factors, such as PTE and breastfeeding, is different among term and preterm infants. Thus, further longitudinal surveys are needed to better clarify the exact relationships. Moreover, in this chapter, we discussed some interdisciplinary works that have been implemented to increase our current knowledge regarding brain development during the first 1000 days of life. There is a wide range of disciplines, including but not limited to pediatrics, neurology, psychiatry, psychology, cellular and molecular biology, radiology, neuroscience, artificial intelligence, bioengineering, and technology, all of which have contributed to new findings in various aspects of brain development and NDDs. Remarkable advances in this field depend on interdisciplinary research, bringing close collaborations between biomedical, behavioral, and clinical scientists in the future.

Conflict of Interest None to declare

Financial Support None

References

1. Cusick SE, Georgieff MK. The Role of Nutrition in Brain Development: The Golden Opportunity of the "First 1000 Days". J Pediatr. 2016;175:16–21. https://doi.org/10.1016/j.jpeds.2016.05.013.
2. Schaub B, Prescott S. The maturation of immune function in pregnancy and early childhood. In: Allergy, immunity and tolerance in early childhood. Amsterdam: Academic Press; 2016. p. 1–17.
3. Derbyshire E, Obeid R. Choline, neurological development and brain function: a systematic review focusing on the first 1000 days. Nutrients. 2020;12(6):1731.
4. Webb SJ, Monk CS, Nelson CA. Mechanisms of postnatal neurobiological development: implications for human development. Dev Neuropsychol. 2001;19(2):147–71.

5. Guimaraes CVA, Dahmoush HM. Fetal brain anatomy. Neuroimaging Clin N Am. 2022;32(3): 663–81.
6. Gilmore JH, Knickmeyer RC, Gao W. Imaging structural and functional brain development in early childhood. Nat Rev Neurosci. 2018;19(3):123–37.
7. Heland S, Fields N, Ellery SJ, Fahey M, Palmer KR. The role of nutrients in human neurodevelopment and their potential to prevent neurodevelopmental adversity. Front Nutr. 2022 [cited 2022 Dec 11]; 9. Available from: https://www.frontiersin.org/articles/10.3389/fnut.2022.992120
8. Theoharides TC, Kavalioti M, Martinotti R. Factors adversely influencing neurodevelopment. J Biol Regul Homeost Agents. 2019;33(6):1663–7.
9. Pervanidou P, Agorastos A, Chrousos GP. Editorial: stress and neurodevelopment. Front Neurosci. 2022;16:898872.
10. Bitta M, Kariuki SM, Abubakar A, Newton CR. Burden of neurodevelopmental disorders in low and middle-income countries: a systematic review and meta-analysis. Wellcome Open Res. 2017;2:121.
11. Brinksma DM, Hoekstra PJ, van den Hoofdakker B, de Bildt A, Buitelaar JK, Hartman CA, et al. Age-dependent role of pre- and perinatal factors in interaction with genes on ADHD symptoms across adolescence. J Psychiatr Res. 2017;90:110–7.
12. Gardener H, Spiegelman D, Buka SL. Prenatal risk factors for autism: comprehensive meta-analysis. Br J Psychiatry. 2009;195(1):7–14.
13. Esposito G, Borelli JL. Investigating genes, environments, and their interactions in the service of informing individualized diagnosis and treatment in developmental disabilities. Res Dev Disabil. 2018;82:1–2.
14. Tautz D. Problems and paradigms: redundancies, development and the flow of information. Bioessays. 1992;14(4):263–6.
15. Graf WD, Kekatpure MV, Kosofsky BE. Prenatal-onset neurodevelopmental disorders secondary to toxins, nutritional deficiencies, and maternal illness. Handb Clin Neurol. 2013;111:143–59.
16. Bishop DV. Genes, cognition, and communication: insights from neurodevelopmental disorders. Ann N Y Acad Sci. 2009;1156(1):1–18.
17. Belsky J, Pluess M. Beyond diathesis stress: differential susceptibility to environmental influences. Psychol Bull. 2009;135(6):885.
18. Linnér A, Almgren M. Epigenetic programming—the important first 1000 days. Acta Paediatr. 2020;109(3):443–52.
19. Isaevska E, Moccia C, Asta F, Cibella F, Gagliardi L, Ronfani L, et al. Exposure to ambient air pollution in the first 1000 days of life and alterations in the DNA methylome and telomere length in children: a systematic review. Environ Res. 2021;193:110504.
20. Pietrobelli A, Agosti M, MeNu Group. Nutrition in the first 1000 days: ten practices to minimize obesity emerging from published science. Int J Environ Res Public Health. 2017;14(12):1491.
21. Esposito G, Azhari A, Borelli JL. Gene × environment interaction in developmental disorders: where do we stand and what's next? Front Psychol. 2018;9:2036.
22. Corrao S, Mallaci Bocchio R, Lo Monaco M, Natoli G, Cavezzi A, Troiani E, et al. Does evidence exist to blunt inflammatory response by nutraceutical supplementation during covid-19 pandemic? An overview of systematic reviews of vitamin d, vitamin c, melatonin, and zinc. Nutrients [Internet]. 2021;13(4):1261. Available from: https://www.embase.com/search/results?subaction=viewrecord&id=L2006934931&from=export
23. Zou R, Boer OD, Felix JF, Muetzel RL, Franken IHA, Cecil CAM, et al. Association of maternal tobacco use during pregnancy with preadolescent brain morphology among offspring. JAMA Netw Open. 2022;5(8):e2224701.
24. Robbins LS, Blanchard CT, Sinkey RG, Harris SL, Tita AT, Harper LM. Prenatal tobacco exposure and childhood neurodevelopment among infants born prematurely. Am J Perinatol. 2021;38(3):218–23.
25. Ylijoki MK, Ekholm E, Ekblad M, Lehtonen L. Prenatal risk factors for adverse developmental outcome in preterm infants—systematic review. Front Psychol [Internet]. 2019 [cited 2022 Dec 2]; 10. Available from: https://www.frontiersin.org/articles/10.3389/fpsyg.2019.00595
26. Polanska K, Krol A, Merecz-Kot D, Ligocka D, Mikolajewska K, Mirabella F, et al. Environmental tobacco smoke exposure during pregnancy and child neurodevelopment. Int J Environ Res Public Health. 2017;14(7):796.
27. Ekblad M, Korkeila J, Lehtonen L. Smoking during pregnancy affects foetal brain development. Acta Paediatr. 2015;104(1):12–8.
28. Zhang Y, Angley M, Qi X, Lu L, D'Alton ME, Kahe K. Maternal electronic cigarette exposure in relation to offspring development: a comprehensive review. Am J Obstet Gynecol MFM [Internet]. 2022 [cited 2022 Dec 2]; 4(5):100659. Available from: https://www.ajogmfm.org/article/S2589-9333(22)00094-5/fulltext.
29. Su X, Zhang S, Lin Q, Wu Y, Yang Y, Yu H, et al. Prenatal exposure to air pollution and neurodevelopmental delay in children: a birth cohort study in Foshan, China. Sci Total Environ. 2022;816:151658.
30. Sunyer J, Dadvand P. Pre-natal brain development as a target for urban air pollution. Basic Clin Pharmacol Toxicol. 2019;125(Suppl 3):81–8.
31. Lin LZ, Zhan XL, Jin CY, Liang JH, Jing J, Dong GH. The epidemiological evidence linking exposure to ambient particulate matter with neurodevelopmental disorders: a systematic review and meta-analysis. Environ Res. 2022;209:112876.
32. Shang L, Yang L, Yang W, Huang L, Qi C, Yang Z, et al. Effects of prenatal exposure to NO2 on children's neurodevelopment: a systematic review and meta-analysis. Environ Sci Pollut Res. 2020;27(20):24786–98.

33. Levy RJ. Carbon monoxide pollution and neurodevelopment: a public health concern. Neurotoxicol Teratol. 2015;49:31–40.
34. Chang G. Maternal substance use: consequences, identification, and interventions. Alcohol Res. 2020;40(2):06.
35. Lee SJ, Bora S, Austin NC, Westerman A, Henderson JMT. Neurodevelopmental outcomes of children born to opioid-dependent mothers: a systematic review and meta-analysis. Acad Pediatr. 2020;20(3):308–18.
36. Mamluk L, Jones T, Ijaz S, Edwards HB, Savović J, Leach V, et al. Evidence of detrimental effects of prenatal alcohol exposure on offspring birthweight and neurodevelopment from a systematic review of quasi-experimental studies. Int J Epidemiol. 2021;49(6):1972–95.
37. Flak AL, Su S, Bertrand J, Denny CH, Kesmodel US, Cogswell ME. The association of mild, moderate, and binge prenatal alcohol exposure and child neuropsychological outcomes: a meta-analysis. Alcohol Clin Exp Res. 2014;38(1):214–26.
38. Lees B, Mewton L, Jacobus J, Valadez EA, Stapinski LA, Teesson M, et al. Association of Prenatal Alcohol Exposure with Psychological, behavioral, and neurodevelopmental outcomes in children from the adolescent brain cognitive development study. Am J Psychiatry. 2020;177(11): 1060–72.
39. Hjorth S, Bromley R, Ystrom E, Lupattelli A, Spigset O, Nordeng H. Use and validity of child neurodevelopment outcome measures in studies on prenatal exposure to psychotropic and analgesic medications—a systematic review. PLoS One. 2019;14(7):e0219778.
40. Bemanalizadeh M, Khoshhali M, Goli P, Abdollahpour I, Kelishadi R. Parental occupational exposure and neurodevelopmental disorders in offspring: a systematic review and meta-analysis. Curr Environ Health Rep. 2022;9(3):406–22.
41. Shah-Kulkarni S, Lee S, Jeong KS, Hong YC, Park H, Ha M, et al. Prenatal exposure to mixtures of heavy metals and neurodevelopment in infants at 6 months. Environ Res. 2020;182:109122.
42. Liu C, Huang L, Huang S, Wei L, Cao D, Zan G, et al. Association of both prenatal and early childhood multiple metals exposure with neurodevelopment in infant: a prospective cohort study. Environ Res. 2022;205:112450.
43. Vrijheid M, Martinez D, Forns J, Guxens M, Julvez J, Ferrer M, et al. Brief report: prenatal exposure to cell phone use and neurodevelopment at 14 months. Epidemiology. 2010;21(2):259–62.
44. Choi KH Ha M, Ha EH, Park H, Kim Y, Hong YC, et al. Neurodevelopment for the first three years following prenatal mobile phone use, radio frequency radiation and lead exposure. Environ Res. 2017;156:810–7.
45. Sudan M, Birks LE, Aurrekoetxea JJ, Ferrero A, Gallastegi M, Guxens M, et al. Maternal cell phone use during pregnancy and child cognition at age 5 years in 3 birth cohorts. Environ Int. 2018;120:155–62.
46. Wu PH, LIN CC, Liao HF, Chen PC. Mobile phone use and early-childhood neurodevelopment at 2 years of age. ISEE Conference Abstracts 2011 Sep 13;2011.
47. Papadopoulou E, Haugen M, Schjølberg S, Magnus P, Brunborg G, Vrijheid M, et al. Maternal cell phone use in early pregnancy and child's language, communication and motor skills at 3 and 5 years: the Norwegian mother and child cohort study (MoBa). BMC Public Health. 2017;17(1):685.
48. Chan JC, Nugent BM, Bale TL. Parental advisory: maternal and paternal stress can impact offspring neurodevelopment. Biol Psychiatry. 2018;83(10):886–94.
49. O'Connor TG, Ciesla AA, Sefair AV, Thornburg LL, Brown AS, Glover V, et al. Maternal prenatal infection and anxiety predict neurodevelopmental outcomes in middle childhood. J Psychopathol Clin Sci. 2022;131(4):422–34.
50. Nazzari S, Frigerio A. The programming role of maternal antenatal inflammation on infants' early neurodevelopment: a review of human studies: special section on "translational and neuroscience studies in affective disorders" section editor, Maria Nobile MD, PhD. J Affect Disord. 2020;263:739–46.
51. Kwon HK, Choi GB, Huh JR. Maternal inflammation and its ramifications on fetal neurodevelopment. Trends Immunol. 2022;43(3):230–44.
52. Doi M, Usui N, Shimada S. Prenatal environment and neurodevelopmental disorders. Front Endocrinol [Internet]. 2022 [cited 2022 Dec 5]; 13. Available from: https://www.frontiersin.org/articles/10.3389/fendo.2022.860110
53. Shook LL, Sullivan EL, Lo JO, Perlis RH, Edlow AG. COVID-19 in pregnancy: implications for fetal brain development. Trends Mol Med. 2022;28(4):319–30.
54. Zimmer A, Youngblood A, Adnane A, Miller BJ, Goldsmith DR. Prenatal exposure to viral infection and neuropsychiatric disorders in offspring: a review of the literature and recommendations for the COVID-19 pandemic. Brain Behav Immun. 2021;91:756–70.
55. Tong L, Kalish BT. The impact of maternal obesity on childhood neurodevelopment. J Perinatol. 2021;41(5):928–39.
56. Edlow AG. Maternal obesity and neurodevelopmental and psychiatric disorders in offspring. Prenat Diagn. 2017;37(1):95–110.
57. Urbonaite G, Knyzeliene A, Bunn FS, Smalskys A, Neniskyte U. The impact of maternal high-fat diet on offspring neurodevelopment. Front Neurosci. 2022;16:909762.
58. Sanchez CE, Barry C, Sabhlok A, Russell K, Majors A, Kollins SH, et al. Maternal pre-pregnancy obesity and child neurodevelopmental outcomes: a meta-analysis. Obes Rev. 2018;19(4):464–84.

59. Batistuzzo A, Ribeiro MO. Clinical and subclinical maternal hypothyroidism and their effects on neurodevelopment, behavior and cognition. Arch Endocrinol Metab. 2020;64(1):89–95.
60. Thompson W, Russell G, Baragwanath G, Matthews J, Vaidya B, Thompson-Coon J. Maternal thyroid hormone insufficiency during pregnancy and risk of neurodevelopmental disorders in offspring: a systematic review and meta-analysis. Clin Endocrinol. 2018;88(4):575–84.
61. He XJ, Dai RX, Tian CQ, Hu CL. Neurodevelopmental outcome at 1 year in offspring of women with gestational diabetes mellitus. Gynecol Endocrinol. 2021;37(1):88–92.
62. Tertti K, Eskola E, Rönnemaa T, Haataja L. Neurodevelopment of two-year-old children exposed to metformin and insulin in gestational diabetes mellitus. J Dev Behav Pediatr. 2015;36(9):752–7.
63. Alemu BT, Olayinka O, Baydoun HA, Hoch M, Akpinar-Elci M. Neonatal hypoglycemia in diabetic mothers: a systematic review. Curr Pediatr Res. 2017;21(1):13.
64. Shah R, Harding J, Brown J, McKinlay C. Neonatal glycaemia and neurodevelopmental outcomes: a systematic review and meta-analysis. Neonatology. 2019;115(2):116–26.
65. Anchan JC, Carr NR, Ahmad KA. Neonatal hypoglycemia: is there a neurodevelopmental impact in early childhood? J Perinatol. 2019;39(1):4–7.
66. Maher GM, O'Keeffe GW, Kearney PM, Kenny LC, Dinan TG, Mattsson M, et al. Association of hypertensive disorders of pregnancy with risk of neurodevelopmental disorders in offspring: a systematic review and meta-analysis. JAMA Psychiatry. 2018;75(8):809–19.
67. Sala ME, Romero MF, Romero A, Fasano MV, Varea AM, Carrera P, et al. Neurodevelopmental assessment of infants born to mothers with hypertensive disorder of pregnancy at six months of age. J Dev Orig Health Dis. 2022;13(2):197–203.
68. Chen Z, Li R, Liu H, Duan J, Yao C, Yang R, et al. Impact of maternal hypertensive disorders on offspring's neurodevelopment: a longitudinal prospective cohort study in China. Pediatr Res. 2020;88(4):668–75.
69. Palatnik A, Mele L, Casey BM, Varner MW, Sorokin Y, Reddy UM, et al. Association between hypertensive disorders of pregnancy and long-term neurodevelopmental outcomes in the offspring. Am J Perinatol. 2022;39(9):921–9.
70. Report of a WHO technical consultation on birth spacing: Geneva, Switzerland 13–15 Jun 2005 [Internet]. 2005 [cited 2022 Dec 27]. Available from: https://apps.who.int/iris/handle/10665/69855
71. Conde-Agudelo A, Rosas-Bermudez A, Norton MH. Birth spacing and risk of autism and other neurodevelopmental disabilities: a systematic review. Pediatrics. 2016;137(5):e20153482.
72. Babatunde OA, Adebamowo SN, Ajayi IO, Adebamowo CA. Neurodevelopmental outcomes of twins compared with singleton children: a systematic review. Twin Res Hum Genet. 2018;21(2):136–45.
73. Yan S, Wang Y, Chen Z, Zhang F. Chorionicity and neurodevelopmental outcomes in twin pregnancy: a systematic review and meta-analysis. J Perinatol. 2023;43(2):133–146. https://doi.org/10.1038/s41372-022-01534-y. Epub 2022 Nov 4. PMID: 36333420.
74. World Health Organization [WHO]. Preterm birth [Internet]. Preterm birth 2022 [cited 2022 Dec 6]. Available from: https://www.who.int/news-room/fact-sheets/detail/preterm-birth
75. Cheong JL, Doyle LW, Burnett AC, Lee KJ, Walsh JM, Potter CR, et al. Association between moderate and late preterm birth and neurodevelopment and social-emotional development at age 2 years. JAMA Pediatr. 2017;171(4):e164805.
76. Sarda SP, Sarri G, Siffel C. Global prevalence of long-term neurodevelopmental impairment following extremely preterm birth: a systematic literature review. J Int Med Res. 2021;49(7):3000605211028026.
77. Marlow N, Ni Y, Lancaster R, Suonpera E, Bernardi M, Fahy A, et al. No change in neurodevelopment at 11 years after extremely preterm birth. Arch Dis Child Fetal Neonatal Ed. 2021;106(4):418–24.
78. Sivanandan S, Gupta S, Sahoo T, Yadav CP, Saxena R, Sikka K, et al. Short-term neurodevelopment and growth outcomes of very and moderate preterm Indian infants. J Perinatol. 2022;42(11):1489–95.
79. Sutton PS, Darmstadt GL. Preterm birth and neurodevelopment: a review of outcomes and recommendations for early identification and cost-effective interventions. J Trop Pediatr. 2013;59(4):258–65.
80. Osuchukwu OO, Reed DJ. Small for gestational age. In: StatPearls [Internet]. Treasure Island: StatPearls Publishing; 2022 [cited 2022 Dec 7]. Available from: http://www.ncbi.nlm.nih.gov/books/NBK563247/
81. Arcangeli T, Thilaganathan B, Hooper R, Khan KS, Bhide A. Neurodevelopmental delay in small babies at term: a systematic review. Ultrasound Obstet Gynecol. 2012;40(3):267–75.
82. Ferguson KK, Sammallahti S, Rosen E, van den Dries M, Pronk A, Spaan S, et al. Fetal growth trajectories among small for gestational age babies and child neurodevelopment. Epidemiology. 2021;32(5):664–71.
83. Sacchi C, Marino C, Nosarti C, Vieno A, Visentin S, Simonelli A. Association of intrauterine growth restriction and small for gestational age status with childhood cognitive outcomes: a systematic review and meta-analysis. JAMA Pediatr. 2020;174(8):772–81.
84. Grissom NM, Reyes TM. Gestational overgrowth and undergrowth affect neurodevelopment: similarities and differences from behavior to epigenetics. Int J Dev Neurosci. 2013;31(6):406–14.

85. Lei X, Zhao D, Huang L, Luo Z, Zhang J, Yu X, et al. Childhood health outcomes in term, large-for-gestational-age babies with different postnatal growth patterns. Am J Epidemiol. 2018;187(3):507–14.
86. Rustogi D, Synnes A, Alshaikh B, Hasan S, Drolet C, Masse E, et al. Neurodevelopmental outcomes of singleton large for gestational age infants <29 weeks' gestation: a retrospective cohort study. J Perinatol. 2021;41(6):1313–21.
87. Zheng X, Li R, Wang L, Yang H, Li L, Cui J, et al. The association of cesarean section with overweight and neurodevelopment of Chinese children aged 1–5 months. Front Pediatr. 2022;10:940422.
88. Zhang T, Sidorchuk A, Sevilla-Cermeño L, Vilaplana-Pérez A, Chang Z, Larsson H, et al. Association of cesarean delivery with risk of neurodevelopmental and psychiatric disorders in the offspring: a systematic review and meta-analysis. JAMA Netw Open. 2019;2(8):e1910236.
89. Chen G, Chiang WL, Chiang TL. Does cesarean delivery increase the occurrence of neurodevelopmental disorders in childhood? Int J Gynaecol Obstet. 2022;158(3):650–6.
90. Zhang T, Brander G, Mantel Ä, Kuja-Halkola R, Stephansson O, Chang Z, et al. Assessment of cesarean delivery and neurodevelopmental and psychiatric disorders in the children of a population-based Swedish birth cohort. JAMA Netw Open. 2021;4(3):e210837.
91. Hassen TA, Chojenta C, Egan N, Loxton D. The association between the five-minute Apgar score and neurodevelopmental outcomes among children aged 8–66 months in Australia. Int J Environ Res Public Health. 2021;18(12):6450.
92. He H, Yu Y, Wang H, Obel CL, Li F, Li J. Five-minute Apgar score and the risk of mental disorders during the first four decades of life: a nationwide registry-based cohort study in Denmark. Front Med (Lausanne). 2021;8:796544.
93. Mohamed NN, Loy SL, Lim PY, Al Mamun A, Jan Mohamed HJ. Early life secondhand smoke exposure assessed by hair nicotine biomarker may reduce children's neurodevelopment at 2years of age. Sci Total Environ. 2018;610–611:147–53.
94. Wang H, Zhang H, Li J, Liao J, Liu J, Hu C, et al. Prenatal and early postnatal exposure to ambient particulate matter and early childhood neurodevelopment: a birth cohort study. Environ Res. 2022;210:112946.
95. Rauh VA, Horton MK, Miller RL, Whyatt RM, Perera F. Neonatology and the environment: impact of early exposure to airborne environmental toxicants on infant and child neurodevelopment. NeoReviews. 2010;11:363–9.
96. Burger M, Hoosain M, Einspieler C, Unger M, Niehaus D. Maternal perinatal mental health and infant and toddler neurodevelopment—evidence from low and middle-income countries. A systematic review. J Affect Disord. 2020;(268):158–72.
97. Nix L, Nixon E, Quigley J, O'Keane V. Perinatal depression and children's developmental outcomes at 2 years postpartum. Early Hum Dev. 2021;156:105346.
98. Siracusano M, Riccioni A, Gialloreti LE, Carloni E, Baratta A, Ferrara M, et al. Maternal perinatal depression and risk of neurodevelopmental disorders in offspring: preliminary results from the SOS MOOD project. Children (Basel). 2021;8(12):1150.
99. Cortés-Albornoz MC, García-Guáqueta DP, Velez-van-Meerbeke A, Talero-Gutiérrez C. Maternal nutrition and neurodevelopment: a scoping review. Nutrients. 2021;13(10):3530.
100. Schwarzenberg SJ, Georgieff MK, Committee on Nutrition. Advocacy for improving nutrition in the first 1000 days to support childhood development and adult health. Pediatrics. 2018;141(2): e20173716.
101. Cohen Kadosh K, Muhardi L, Parikh P, Basso M, Jan Mohamed HJ, Prawitasari T, et al. Nutritional support of neurodevelopment and cognitive function in infants and young children—an update and novel insights. Nutrients. 2021;13(1):199.
102. Nevins JEH, Donovan SM, Snetselaar L, Dewey KG, Novotny R, Stang J, et al. Omega-3 fatty acid dietary supplements consumed during pregnancy and lactation and child neurodevelopment: a systematic review. J Nutr. 2021;151(11):3483–94.
103. Duttaroy AK. Special issue "maternal DHA impact on child neurodevelopment.". Nutrients. 2021;13(7):2209.
104. Basak S, Mallick R, Banerjee A, Pathak S, Duttaroy AK. Maternal supply of both arachidonic and docosahexaenoic acids is required for optimal neurodevelopment. Nutrients. 2021;13(6):2061.
105. Iglesias L, Canals J, Arija V. Effects of prenatal iron status on child neurodevelopment and behavior: a systematic review. Crit Rev Food Sci Nutr. 2018;58(10):1604–14.
106. McCann S, Perapoch Amadó M, Moore SE. The role of iron in brain development: a systematic review. Nutrients. 2020;12(7):2001.
107. Jayasinghe C, Polson R, van Woerden HC, Wilson P. The effect of universal maternal antenatal iron supplementation on neurodevelopment in offspring: a systematic review and meta-analysis. BMC Pediatr. 2018;18(1):150.
108. Al Rubaye H, Adamson CC, Jadavji NM. The role of maternal diet on offspring gut microbiota development: a review. J Neurosci Res. 2021;99(1):284–93.
109. Tognini P. Gut microbiota: a potential regulator of neurodevelopment. Front Cell Neurosci. 2017; 11:25.
110. Morales-Suárez-Varela M, Clemente-Bosch E, Peraita-Costa I, Llopis-Morales A, Martínez I, Llopis-González A. Maternal physical activity during pregnancy and the effect on the mother and newborn: a systematic review. J Phys Act Health. 2021;18(1):130–47.

111. Ribeiro MM, Andrade A, Nunes I. Physical exercise in pregnancy: benefits, risks and prescription. J Perinat Med. 2022;50(1):4–17.
112. Niño Cruz GI, Ramirez Varela A, da Silva ICM, Hallal PC, Santos IS. Physical activity during pregnancy and offspring neurodevelopment: a systematic review. Paediatr Perinat Epidemiol. 2018;32(4):369–79.
113. Álvarez-Bueno C, Cavero-Redondo I, Sánchez-López M, Garrido-Miguel M, Martínez-Hortelano JA, Martínez-Vizcaíno V. Pregnancy leisure physical activity and children's neurodevelopment: a narrative review. BJOG. 2018;125(10):1235–42.
114. World Health Organization [WHO]. Exclusive breastfeeding for six months best for babies everywhere [Internet]. 2011 [cited 2022 Dec 8]. Available from: https://www.who.int/news/item/15-01-2011-exclusive-breastfeeding-for-six-months-best-for-babies-everywhere
115. Horta BL, Loret de Mola C, Victora CG. Breastfeeding and intelligence: a systematic review and meta-analysis. Acta Paediatr. 2015;104(467):14–9.
116. Bar S, Milanaik R, Adesman A. Long-term neurodevelopmental benefits of breastfeeding. Curr Opin Pediatr. 2016;28(4):559–66.
117. Hou L, Li X, Yan P, Li Y, Wu Y, Yang Q, et al. Impact of the duration of breastfeeding on the intelligence of children: a systematic review with network meta-analysis. Breastfeed Med. 2021;16(9):687–96.
118. Lechner BE, Vohr BR. Neurodevelopmental outcomes of preterm infants fed human milk: a systematic review. Clin Perinatol. 2017;44(1):69–83.
119. Miller J, Tonkin E, Damarell RA, McPhee AJ, Suganuma M, Suganuma H, et al. A systematic review and meta-analysis of human milk feeding and morbidity in very low birth weight infants. Nutrients. 2018;10(6):707.
120. Yang Q, Yang J, Zheng L, Song W, Yi L. Impact of home parenting environment on cognitive and psychomotor development in children under 5 years old: a meta-analysis. Front Pediatr [Internet]. 2021 [cited 2022 Dec 9]; 9. Available from: https://www.frontiersin.org/articles/10.3389/fped.2021.658094
121. Bush NR, Wakschlag LS, LeWinn KZ, Hertz-Picciotto I, Nozadi SS, Pieper S, et al. Family environment, neurodevelopmental risk, and the environmental influences on child health outcomes (ECHO) initiative: looking back and moving forward. Front Psych. 2020;11:547.
122. Jeong J, Franchett EE, Oliveira CVR d, Rehmani K, Yousafzai AK. Parenting interventions to promote early child development in the first three years of life: a global systematic review and meta-analysis. PLoS Med. 2021;18(5):e1003602.
123. Bemanalizadeh M, Badihian N, Khoshhali M, Badihian S, Hosseini N, Purpirali M, et al. Effect of parenting intervention through "Care for Child Development Guideline" on early child development and behaviors: a randomized controlled trial. BMC Pediatr. 2022;22(1):690.
124. Eisenberg L, Pellmar TC. Bridging disciplines in the brain, behavioral, and clinical sciences. Washington, DC: National Academies Press; 2000.
125. Pellmar TC, Eisenberg L. The potential of interdisciplinary research to solve problems in the brain, behavioral, and clinical sciences. In: Bridging disciplines in the brain, behavioral, and clinical sciences. Washington, DC: National Academies Press; 2000.
126. Leoni M, Corti S, Cavagnola R. Third generation behavioural therapy for neurodevelopmental disorders: review and trajectories. Adv Ment Health Intellect Disabil. 2015;9(5):265–74.
127. Tărlungeanu DC, Novarino G. Genomics in neurodevelopmental disorders: an avenue to personalized medicine. Exp Mol Med. 2018;50(8):1–7.
128. Arabi F, Mansouri V, Ahmadbeigi N. Gene therapy clinical trials, where do we go? An overview. Biomed Pharmacother. 2022;153:113324.
129. Global Research on Developmental Disabilities Collaborators. Developmental disabilities among children younger than 5 years in 195 countries and territories, 1990–2016: a systematic analysis for the Global Burden of Disease Study 2016. Lancet Glob Health. 2018;6(10):e1100–21.
130. Shashi V, McConkie-Rosell A, Rosell B, Schoch K, Vellore K, McDonald M, et al. The utility of the traditional medical genetics diagnostic evaluation in the context of next-generation sequencing for undiagnosed genetic disorders. Genet Med. 2014;16(2):176–82.
131. de la Torre-Ubieta L, Won H, Stein JL, Geschwind DH. Advancing the understanding of autism disease mechanisms through genetics. Nat Med. 2016;22(4):345–61.
132. Bruining H, Eijkemans MJ, Kas MJ, Curran SR, Vorstman JA, Bolton PF. Behavioral signatures related to genetic disorders in autism. Mol Autism. 2014;5(1):1–12.
133. Volkmar F, Siegel M, Woodbury-Smith M, King B, McCracken J, State M. Practice parameter for the assessment and treatment of children and adolescents with autism spectrum disorder. J Am Acad Child Adolesc Psychiatry. 2014;53(2):237–57.
134. Bernardet M, Crusio WE. Fmr1 KO mice as a possible model of autistic features. ScientificWorldJournal. 2006;6:1164–76.
135. Grayton HM, Missler M, Collier DA, Fernandes C. Altered social behaviours in neurexin 1α knockout mice resemble core symptoms in neurodevelopmental disorders. PLoS One. 2013;8(6):e67114.
136. Wang X, McCoy PA, Rodriguiz RM, Pan Y, Je HS, Roberts AC, et al. Synaptic dysfunction and abnormal behaviors in mice lacking major isoforms of Shank3. Hum Mol Genet. 2011;20(15):3093–108.
137. Baudouin SJ, Gaudias J, Gerharz S, Hatstatt L, Zhou K, Punnakkal P, et al. Shared synaptic pathophysiology in syndromic and nonsyndromic rodent models of autism. Science. 2012;338(6103):128–32.
138. Han S, Tai C, Westenbroek RE, Yu FH, Cheah CS, Potter GB, et al. Autistic-like behaviour in Scn1a+/−

mice and rescue by enhanced GABA-mediated neurotransmission. Nature. 2012;489(7416):385–90.
139. Hulbert SW, Jiang YH. Monogenic mouse models of autism spectrum disorders: common mechanisms and missing links. Neuroscience. 2016;321:3–23.
140. Watson KK, Platt ML. Of mice and monkeys: using non-human primate models to bridge mouse- and human-based investigations of autism spectrum disorders. J Neurodev Disord. 2012;4(1):21.
141. McCammon JM, Sive H. Addressing the genetics of human mental health disorders in model organisms. Annu Rev Genomics Hum Genet. 2015;16:173–97
142. Marchetto MC, Carromeu C, Acab A, Yu D, Yeo GW, Mu Y, et al. A model for neural development and treatment of Rett syndrome using human induced pluripotent stem cells. Cell. 2010;143(4):527–39.
143. Chamberlain SJ, Chen PF, Ng KY, Bourgois-Rocha F, Lemtiri-Chlieh F, Levine ES, et al. Induced pluripotent stem cell models of the genomic imprinting disorders Angelman and Prader-Willi syndromes. Proc Natl Acad Sci USA. 2010;107(41):17668–73.
144. Shcheglovitov A, Shcheglovitova O, Yazawa M, Portmann T, Shu R, Sebastiano V, et al. SHANK3 and IGF1 restore synaptic deficits in neurons from 22q13 deletion syndrome patients. Nature. 2013;503(7475):267–71.
145. Urbach A, Bar-Nur O, Daley GQ, Benvenisty N. Differential modeling of fragile X syndrome by human embryonic stem cells and induced pluripotent stem cells. Cell Stem Cell. 2010;6(5):407–11.
146. Jiao J, Yang Y, Shi Y, Chen J, Gao R, Fan Y, et al. Modeling Dravet syndrome using induced pluripotent stem cells (iPSCs) and directly converted neurons. Hum Mol Genet. 2013;22(21):4241–52.
147. Paşca SP, Portmann T, Voineagu I, Yazawa M, Shcheglovitov A, Paşca AM, et al. Using iPSC-derived neurons to uncover cellular phenotypes associated with Timothy syndrome. Nat Med. 2011;17(12):1657–62.
148. Paşca AM, Sloan SA, Clarke LE, Tian Y, Makinson CD, Huber N, et al. Functional cortical neurons and astrocytes from human pluripotent stem cells in 3D culture. Nat Methods. 2015;12(7):671–8.
149. Hattori N. Cerebral organoids model human brain development and microcephaly. Mov Disord. 2014;29(2):185.
150. Campolo D, Laschi C, Keller F, Guglielmelli E. A mechatronic platform for early diagnosis of neurodevelopmental disorders. Adv Robot. 2007;21(10):1131–50.
151. Campolo D, Molteni M, Guglielmelli E, Keller F, Laschi C, Dario P. Towards development of biomechatronic tools for early diagnosis of neurodevelopmental disorders. Conf Proc IEEE Eng Med Biol Soc. 2006:3242–5.
152. Campolo D, Taffoni F, Schiavone G, Laschi C, Keller F, Guglielmelli E. A novel technological approach towards the early diagnosis of neurodevelopmental disorders. Annu Int Conf IEEE Eng Med Biol Soc. 2008:4875–8.
153. Valentine AZ, Hall SS, Young E, Brown BJ, Groom MJ, Hollis C, et al. Implementation of telehealth services to assess, monitor, and treat neurodevelopmental disorders: systematic review. J Med Internet Res. 2021;23(1):e22619.
154. Bemanalizadeh M, Yazdi M, Yaghini O, Kelishadi R. A meta-analysis on the effect of telemedicine on the management of attention deficit and hyperactivity disorder in children and adolescents. J Telemed Telecare. 2021;30(1):31–43. https://doi.org/10.1177/1357633X211045186.
155. Khan K, Hall CL, Davies EB, Hollis C, Glazebrook C. The effectiveness of web-based interventions delivered to children and Young people with neurodevelopmental disorders: systematic review and meta-analysis. J Med Internet Res. 2019;21(11):e13478.
156. Knutsen J, Wolfe A, Burke BL, Hepburn S, Lindgren S, Coury D, et al. A systematic review of telemedicine in autism spectrum disorders. Rev J Autism Dev Disord. 2016;3(4):330–44.
157. Yerys BE, Pennington BF. How do we establish a biological marker for a behaviorally defined disorder? Autism as a test case. Wiley Online Library; 2011.
158. Bednarz HM, Kana RK. Advances, challenges, and promises in pediatric neuroimaging of neurodevelopmental disorders. Neurosci Biobehav Rev. 2018;90:50–69.
159. Gupta C, Chandrashekar P, Jin T, He C, Khullar S, Chang Q, et al. Bringing machine learning to research on intellectual and developmental disabilities: taking inspiration from neurological diseases. J Neurodev Disord. 2022;14(1):28.
160. Guilé JM, Tissot C, Boissel L. Interdisciplinary assessment. In: Handbook of clinical neurology. Elsevier; 2020. p. 173–81.
161. Insel TR, Cuthbert BN. Brain disorders? Precisely. Science. 2015;348(6234):499–500.
162. Kapur S, Phillips AG, Insel TR. Why has it taken so long for biological psychiatry to develop clinical tests and what to do about it? Mol Psychiatry. 2012;17(12):1174–9.
163. Lee JW, Devanarayan V, Barrett YC, Weiner R, Allinson J, Fountain S, et al. Fit-for-purpose method development and validation for successful biomarker measurement. Pharm Res. 2006;23(2):312–28.
164. Loth E, Evans DW. Converting tests of fundamental social, cognitive, and affective processes into clinically useful bio-behavioral markers for neurodevelopmental conditions. Wiley Interdiscip Rev Cogn Sci. 2019;10(5):e1499.

Etiology and Translational Research in Neurodevelopmental Disorders

Fei Li, Mingyu Xu, Juehua Yu, and Miao Cao

Abstract

Neurodevelopmental disorders (NDDs) cover a range of clinical conditions that affect the functioning of the brain, disturbing the coordinated processes essential for healthy brain development. This disturbance gives rise to developmental deficits, leading to enduring challenges in personal, social, academic, or occupational performance. Generally, these disorders become evident early in development and can range from specific difficulties in learning or executive function control to broader impairments in social skills or intelligence.

In spite of thorough research, the clinical management of NDDs faces significant challenges due to a lack of comprehensive understanding of their origins, not to mention the successful translation of this knowledge into effective therapies. Advancements in creating new medications for NDDs have been constrained, underscoring the necessity for a more resilient strategy. To improve pharmacotherapy and drug discovery, it is vital to concentrate on addressing numerous endophenotypes, carefully analyzing genetic and environmental influences, and taking into account the potential developmental and trophic functions of brain neurostructure and neuroimaging.

Tackling the intricate challenges linked to NDDs requires a more profound comprehension of their pathobiology, encouraging innovative cross-disciplinary research that goes beyond conventional methods and concepts. By giving prominence to these crucial aspects, we can pave the way for progress in the treatment and handling of neurodevelopmental disorders.

Keywords

Neurodevelopmental disorders · Autism spectrum disorder · Developmental delay/intellectual disability · Attention deficit hyperactivity disorder · Genes · Neuroimaging

F. Li (✉) · M. Xu
Department of Developmental and Behavioral Pediatric and Child Primary Care,
Brain and Behavioral Research Unit of Shanghai Institute for Pediatric Research and Ministry of Education-Shanghai Key Laboratory for Children's Environmental Health, Xinhua Hospital,
Shanghai Jiao Tong University School of Medicine, Shanghai, China
e-mail: feili@shsmu.edu.cn

J. Yu
Department of Developmental and Behavioral Pediatric and Child Primary Care, Xinhua Hospital, Shanghai Jiao Tong University School of Medicine, Shanghai, China

M. Cao
Institute of Science and Technology for Brain-Inspired Intelligence, Key Laboratory of Computational Neuroscience and Brain-Inspired Intelligence, Ministry of Education, Fudan University, Shanghai, China

© The Author(s), under exclusive license to Springer Nature Singapore Pte Ltd. 2024

R. Kelishadi (ed.), *Nature, Nurture, and Neurodevelopment*,
https://doi.org/10.1007/978-981-97-7433-3_2

1 Multifactorial Etiology of Neurodevelopmental Disorders

Neurodevelopmental disorders (NDDs) constitute a set of clinical conditions impacting brain function, disrupting the synchronized processes essential for brain development. This disruption gives rise to developmental deficits, leading to enduring impairments in personal, social, academic, or occupational functioning throughout one's lifetime. Typically emerging early in the developmental stage, these disorders can vary from highly specific challenges related to learning or executive function control to broader impairments affecting social skills or intelligence. NDDs have a multifactorial etiology, including autism spectrum disorder (ASD), developmental delay/intellectual disability (DD/ID), and attention deficit hyperactivity disorder (ADHD). Several studies suggest that these disorders share common molecular pathways, contributing to the diverse clinical manifestations that characterize NDDs. Comorbidity of two or more of these disorders is frequently observed, such as a combination of ASD and DD/ID, or ASD and ADHD. Therefore, identifying shared pathogenic mechanisms among different types of NDDs will help elucidate comorbidity and, ultimately, lead to precise diagnosis and effective treatment.

1.1 Genetics of NDDs

NDDs are caused by the disruption of essential neurodevelopmental processes, and many genes and mutations are associated with NDDs, pointing to an extremely complex array of genetic variations that confer higher vulnerability to different kinds of adverse environmental factors during pregnancy, such as maternal infections, obstetric complications, and exposure to drugs or neurotoxicants. Crucial for comprehending the molecular mechanisms triggering the onset of these disorders, the identification of potential genetic causes in NDDs is vital. It also aids in establishing a genotype–phenotype correlation, facilitating the monitoring of disorder progression and the anticipation of future complications. However, the identification of a potential underlying mutational event, known as molecular diagnosis, is a challenging task that must overcome the heterogeneity of this complex array of genetic variations. Even with the identification of numerous genes causing NDDs, a considerable number of individuals with these disorders still lack a molecular diagnosis. Furthermore, studies on phenotype–genotype correlation reveal significant variations in the number and severity of clinical signs among patients sharing the same genetic origin. Hence, the presence of unaccounted heritability and variability in phenotypes suggests a multifactorial and/or polygenic nature of NDDs. Identifying genes responsible for NDDs is vital for precise genetic counseling and effective patient management, serving as a fundamental initial stride toward gaining a deeper insight into the molecular pathways influenced by these disorders.

1.2 Essential Genetic Findings in Neurodevelopmental Disorders

Familial NDDs serve as a valuable model to explore the interplay of genetic and nongenetic factors in the pathogenesis of these disorders within a shared genetic context. Research has focused on monozygotic twins exhibiting different phenotypes or on pedigrees where multiple affected offspring show incomplete penetrance and phenotypic variability. This avenue of study holds significant promise not only for uncovering the molecular origins of the disease but also for identifying risk and protective factors, leading to more precise genotype–phenotype correlations. The investigation of inherited NDDs has unveiled that the phenotypic outcome primarily hinges on two key principles: gene vulnerability and mutational load. Gene vulnerability pertains to a specific gene's capacity to withstand disruptive variants, and certain genes linked to NDDs are haploinsufficient genes known for their remarkable dosage sensitivity. These particular genes belong to the category of highly susceptible genes, and mutations affecting them are associated with a significant risk of disease. Examples of highly susceptible genes include CACNA1A and SCN8A. Disruption of either of these genes markedly raises the probability of triggering a disease phenotype, even in the absence of additional causative factors, leading to monogenic forms of NDDs. Mutations affecting these genes typically face strong negative selective pressure, as evidenced by population studies that indicate a lower occurrence of disruptive variants in vulnerable genes compared to other genomic locations. In essence, mutations in highly susceptible genes can be characterized as rare variants linked to substantial disease risk and high penetrance.

On the opposite end of the vulnerability spectrum are genes less prone to disruptive mutations. Variants within these genes face less negative selective pressure and are frequently inherited across generations within families. Since individual disruptive events in nonvulnerable genes are not inherently disease-causing, they fall into the realm of common variants associated with a low risk of disease. Nevertheless, recent studies indicate that a significant portion of NDDs with a polygenic nature can be traced back to common genetic variants. In these instances, the combined impacts of these mutational events may result in a disease phenotype. However, the phenotypic outcome in such scenarios is determined not only by the cumulative effects of individual mutations but also by the physical and/or functional interactions between the affected genes. Epistatic interactions and dosage sensitivity are closely tied to the concept of mutational load, suggesting that the penetrance and complexity of a disease phenotype are affected by the number of disruptive events. Presently, alterations in over 1000 genes have been associated with ASD, although many of these associations await validation. While specific genetic variations, such as copy number variation (CNV), single nucleotide polymorphism (SNP), and gene mutations, have been implicated, their individual effects are generally modest. Instead, alterations in multiple genes can have a cumulative impact or interact with environmental risk factors, such as parental age and birth complications, to elevate the risk of developing ASD. The contribution of nongenetic factors to ASD has been estimated as 40–60%. In 10–20% of individuals with ASD, a specific gene mutation or chromosomal abnormality is the major causative factor of the syndrome. For example, mutation in the methyl-CpG binding protein 2 (MeCP2) gene is known to cause Rett syndrome (no longer classified with ASD in the DSM-5). In addition to autism and intellectual disability, Rett syndrome is characterized by distinctive facial features and an array of other symptoms. Other rare genes associated with ASD include AT-rich interactive domain-containing protein 1B (ARID1B), ASH1-like histone lysine methyltransferase (ASH1L), chromodomain helicase DNA-binding protein 2 (CHD2), CHD8, dual specificity tyrosine phosphorylation regulated kinase 1A (DYRK1A), fragile X mental retardation 1 (FMR1), Pogo transposable element-derived with ZNF domain (POGZ), SH3 and multiple ankyrin repeat domains 3 (SHANK3), and synaptic Ras GTPase-activating protein 1 (SYNGAP1). Recent studies on gene function have demonstrated the involvement of these genes in various aspects of brain development, encompassing the generation, growth, and organization of neurons. The specific mechanisms through which alterations in these genes contribute to the development of ASD are not yet fully understood. Neuronal and brain abnormalities in the frontal and temporal lobe cortex—which are mainly involved in emotions, social behavior, and language development—occur in ASD and are thought to underlie impairments in socialization, communication, and cognitive functioning. As an illustration, mutations that result in the loss of function in a single allele of sodium channels CACNA1A and SCN8A are frequently linked to a range of clinical characteristics, such as movement disorders, intellectual disability (ID), ASD, and benign familial infantile seizures. Consistent with the criteria of dosage sensitivity and mutational load mentioned earlier, inherited germline mutations affecting both alleles of CACNA1A and SCN8A are connected to more severe phenotypes when compared to mutations affecting a single allele.

Variants, both primary and secondary, can manifest at distinct genomic loci, thereby expanding beyond the conventional two-hit hypothesis. Numerous studies have illuminated the substantial impact of multiple molecular diagnoses in the realm of NDDs. Consistent with the concept of mutational load, analyses correlating genotype and phenotype have demonstrated that individuals with mutations in multiple genes are more prone to being affected. Moreover, the number of disruptive events is positively associated with both the quantity and severity of observed clinical signs. Illustratively, the contribution of diverse mutational events was recently scrutinized in two families characterized by intrafamilial clinical variability. In both instances, additional clinical features in the probands were attributed to mutations at distinct loci, which were absent in the less severely affected siblings.

The accumulated burden of common genetic variants could serve as the initial impact, rendering the genetic background more susceptible to subsequent pathological occurrences. Recent findings indicate that the collective impact of common variants in families with a history of NDDs is positively correlated with a genetic predisposition to lower educational attainment and developmental delay/intellectual disability (DD/ID). Consequently, the available data in current literature suggest that purely monogenic forms of NDDs are more of an exception than a rule. The majority of NDD cases likely possess a multifactorial and/or polygenic nature, underscoring the extensive heterogeneity of these disorders at both clinical and molecular levels. It is worth noting that the clinical outcome might also be influenced by various nongenetic factors at different levels, although a discussion of environmental factors exceeds the scope of the current review.

Advances in genetics and genomics technology, large cohort studies, and the establishment of large databases have enabled the identification and validation of causal genes in ASD and other neurodevelopmental disorders sharing phenotypic and behavioral features such as ADHD and developmental/intellectual disability. The integration of next-generation sequencing (NGS) technologies into the diagnostic framework for NDDs has significantly elevated the proportion of patients receiving a molecular diagnosis. Unraveling the genetic basis of the disease holds substantial implications for genetic counseling and patient management, enabling a more precise assessment of recurrence risk and the ability to anticipate future medical complications. Progress in the field of genetics has not only paved the way for functional genomic studies focused on comprehending the pathogenic mechanisms linked to reported mutations but has also provided valuable insights into the biological pathways crucial for NDD onset. Despite the heterogeneous behavioral and cognitive manifestations of NDDs, which can complicate the identification of genetic risk factors, this avenue of research has brought clarity to some of the underlying biological pathways. Recognizing these networks presents an opportunity to navigate the complexities associated with extensive genetic variability and to formulate targeted therapeutic approaches.

1.3 Principal Molecular Pathways Affected in NDDs

Over the past decade, functional studies have revealed that both rare and common variants associated with NDDs predominantly impact genes involved in a select few conserved pathways. The establishment of "The Psychiatric Cell Map Initiative" a few years ago aimed to unravel the molecular pathophysiology of NDDs and delineate key biological pathways across temporal and spatial axes. This initiative highlighted that both common and rare variants lead to disruptions in homeostatic equilibrium at various levels—whether at the cellular, circuit, or whole brain level. This review categorizes numerous genetic variants based on their effects on specific functional molecular pathways. The pathways scrutinized in detail encompass: (i) protein synthesis; (ii) transcriptional or epigenetic regulation; and (iii) synaptic signaling. Significantly, many mutations seem to be interconnected in a multipathway loop. In this context, second-hit mutations in highly vulnerable genes or the accumulation of common variants can impact the entire loop, underscoring the critical role of the aforementioned pathways in the onset of the disorder.

1.4 Environmental Risk Factors for NDDs

Neurodevelopmental disorders (NDDs) manifest as changes in brain functioning, structure, and maturation, leading to challenges in cognitive and adaptive abilities. In the United States, the prevalence of NDDs is estimated to be 10–15% of all births. Currently, the most frequently diagnosed NDDs include autism spectrum disorder (ASD), attention-deficit/hyperactivity disorder (ADHD), and intellectual disability (ID). Various studies have pinpointed a diverse array of potential environmental risk factors associated with NDDs. A recent umbrella review systematically and quantitatively appraised environmental risk factors for ASD [1]. In this investigation, elements of maternal metabolic syndrome (including chronic hypertension, gestational hypertension, preeclampsia, and overweight) were linked to an augmented risk of ASD in offspring, all categorized as compelling evidence. Elevated maternal age, particularly in the comparison of age groups 35 years or older versus 25–29 years, was also correlated with an increased risk of ASD, deemed as compelling evidence. Additionally, an advanced paternal age was associated with a heightened incidence of ASD. Simultaneously, compelling evidence suggested a heightened risk of ASD in offspring when mothers used selective serotonin reuptake inhibitors (SSRIs), an antidepressant, during pregnancy compared to unexposed groups. However, considering that the presence of a maternal psychiatric condition independently increases the risk of ASD, the connection between SSRI use during pregnancy and ASD requires cautious interpretation and should be further confirmed through well-designed future studies. Maternal autoimmune disease was linked to an elevated risk of ASD, categorized as having a highly suggestive association. Similarly, a family history of psoriasis, rheumatoid arthritis, type 1 diabetes, or any autoimmune disease was associated with an increased risk of ASD, marked as suggestive evidence. To summarize, this umbrella review revealed that associations with convincing evidence (class I) predominantly involved maternal factors, such as age, features of metabolic syndrome, and the use of SSRIs. The associations of ASD with higher paternal age, maternal autoimmune disease, and acetaminophen use during pregnancy were graded as highly suggestive evidence (class II), partially due to the presence of small study effects and substantial heterogeneity. Following several sensitivity subset analyses, only two associations—maternal overweight before or during pregnancy and SSRI use during pregnancy—remained as either convincing or highly suggestive evidence.

ADHD is linked to a range of environmental exposures. Various prenatal and perinatal risk factors, including prematurity and low birth weight, consistently show associations with ADHD, and family studies suggest that these effects cannot be solely attributed to genetic factors. Intrauterine exposure to tobacco, maternal stress, and pregnancy-related obesity is also correlated with ADHD, although these connections may be influenced, in part, by confounding genetic factors. The evidence remains inconclusive regarding the impact of intrauterine exposure to alcohol, drugs, and birth-related complications. Postnatal factors and social determinants also play a role in ADHD. Experimental findings suggest that exposure to artificial food colorings and flavorings can heighten the severity of ADHD symptoms, albeit with modest effects. Associations between ADHD and parenting style have been noted, but they may result from an evocative gene–environment correlation. In this context, a child's behavior triggers harsh and unsupportive parenting, leading to an escalation of problems and the development of coercive cycles within families. As most of these associations rely on observational rather than experimental study designs, establishing causality remains a challenge. Innovative approaches, such as Mendelian randomization, are now being employed to explore causal interpretations of these effects.

Various environmental risk factors during prenatal, perinatal, and postnatal stages may contribute to the development of intellectual disabilities (ID). Prenatal exposure to alcohol is associated with neurocognitive problems. Intrauterine infections, such as TORCH infections, may manifest as developmental disabilities, growth defects, and abnormalities in head size, such as microcephaly (e.g., cytomegalovirus) or macrocephaly (e.g., toxoplasmosis associated with hydrocephalus), potentially leading to ID. Significant hypoxic-ischemic injury in neonates may result in substantial global developmental delays, microcephaly or poor head growth, and seizures. Children with severe hypoxic-ischemic encephalopathy may eventually be diagnosed with ID. Prematurity, especially in cases of younger gestational age and more complicated courses, also places a child at risk for intellectual impairment and developmental disorders. Exposure to environmental lead during early life has been linked to neurocognitive impairment, with lead toxicity associated with a decline of 1 to 2 IQ points (measured at 5 years or older) for every 10-point increase in lead levels.

1.5 Abnormal Brain Development in NDD Patients Revealed by Neuroimaging Research

In recent decades, in vivo neuroimaging techniques, such as functional near-infrared spectroscopy, electroencephalography/magnetoencephalography, and structural, diffusion, and functional MRI, have provided a fascinating glimpse into the human brain. These advancements have led to an unparalleled comprehension of the atypical developmental mechanisms underlying neurodevelopmental disorders, including ASD, ADHD, and ID.

Abnormal Brain Development in ASD Neuroimaging studies revealed altered brain development in ASD compared with healthy populations [53]. Specifically, the crossed early developmental curves between autism (or autism-risk) populations and healthy controls were observed in both the brain structure and function. A meta-analysis of studies on brain size in autism revealed a slight reduction in brain size at birth, a significant increase within the first year of life, but subsequently reaching a plateau. By adulthood, the majority of cases fell within the normal range. Notably, the brain size of the youngest ages (2–5) among all age groups (young child, older child, adult) showed the most significant deviation from the norm was observed in another study. This study utilized sMRI data from 21 medication-naive individuals with high-functioning ASD aged between 7 and 15 years, along with 21 matched controls. The findings indicated a notable increase of 6% in intracranium, total brain, cerebral gray matter, and cerebellum in ASD patients, along with an increase of over 40% in lateral and third ventricles compared to the control group. Diffusion MRI studies showed that the white matter structure of ASD patients aged 3 years old had more short-range U-fibers in the frontal lobe compared to the control brain. Analysis of structural connectivity using diffusion imaging data revealed divergent developmental trajectories between subjects with autism and healthy controls from 2 to 7 years old, with the trajectories intersecting around the age of 4 years. A study utilizing functional near-infrared spectroscopy reported that infants at high risk for autism demonstrated elevated overall functional connectivity strength at 3 months old in comparison to subjects at low risk. The disparities lessened as development progressed and underwent a reversal in the 12-month-old infants. Specifically, functional connectivity decreased in the high-risk autism group when compared to the low-risk group.

Models of abnormal development for ASD have been subsquently proposed, such as immature or ectopic wiring. Brain disruptions observed through neuroimaging seemed to result from developmental hypoplasia rather than shrinkage or deterioration after reaching full development. Current studies are conducted to test these hypotheses with new statistical models. With a normative model, Bethlehem et al. separated a subgroup of ASD patients characterized by abnormal cortical thickness development, proving the advantages of individualized analysis for understanding the heterogeneity of ASD. Addtionally, by calculating extreme deviations (<1%) in brain GMV asymmetry in ASD patients, Floris et al. found that the percentage of voxels with extreme deviation in more than 2% of the subjects was approximately 1%.

1.6 Abnormal Brain Development in ADHD and DD

More than a decade of neuroimaging studies has characterized the structural and functional neural correlates of ADHD and demonstrated that ADHD is not only associated with discrete brain abnormalities but is also involved in disruptions of distributed brain networks. For example, Gao et al. (2019) uncovered impaired interactions among the default, frontoparietal, affective, attention, and somatosensory networks in individuals with ADHD. The abnormal interactions between the DMN and other networks suggested that the DMN, which is hypothesized to underlie mind-wandering, may interfere with attentional and executive networks' capacity to stay focused. Notably, the two meta-analyses yielded quite disparate results: the voxel-based analysis (VBA) meta-analysis identified increased fractional anisotropy (FA) in the left cingulate, anterior corpus callosum (CC), and left inferior fronto-occipital fasciculus (IFOF) in ADHD, along with decreased FA in the anterior cingulate and bilateral orbitofrontal white matter. However, the tract-based spatial statistics (TBSS) meta-analysis only revealed decreased FA in the isthmus of the CC, posterior midbody of the CC, right IFOF, left inferior longitudinal fasciculus (ILF), and right superior longitudinal fasciculus (SLF). Given concerns about head motion in ADHD, they also qualitatively assessed the impact of head motion on TBSS studies. As a result, they found that a substantial number of studies did not investigate group differences in head motion, and nearly half of the studies that found no distinction in head motion also reported no significant difference in FA. This study, on the one hand, provided evidence of the involvement of the CC in the pathophysiology of ADHD, suggesting atypical interhemispheric connections in ADHD. On the other hand, it demonstrated that the results of DTI meta-analyses may be influenced by group differences in head motion. Therefore, future DTI studies should control and correct head motion to avoid the influence of group differences in head motion.

Global topological analyses have revealed reductions in both global integration and local clustering in both structural and functional connectomes, and these findings indicate a regular shift in organizational properties in ADHD patients. Furthermore, an EEG study reported that, compared with controls, ADHD patients exhibited a lower global brain network, which was improved (increased global network efficiency) after effective neurofeedback therapies. Notably, no functional global topological alterations were observed in ADHD adults, although regional disturbances remained. Disrupted hubs have also been found in both functional and structural networks, supporting the conclusion that global integration is disturbed in ADHD.

One popular hypothesis is that wiring abnormalities in patients' brains may arise from a dysplastic developmental lesion. In ADHD, the significantly decreased functional connectivity pattern in the default mode network greatly overlaps with that showing developmental dynamics in healthy populations, which indicates that the ADHD population is associated with developmental delay. Recently, a study tested the hypothesis with subjects aged from 7 to 21 years old to compare the development curves of ADHD and healthy controls. The authors found that ADHD children exhibited a lag in the maturation of both within and between subnetwork functional connections. Still, further studies including both structural evidence and comprehensive connectome organization analysis are needed to corroborate the maturation-lag theory.

Few neuroimaging studies have been conducted on the DD population. But it has been revealed that the neural mechanisms of DD/ID in childhood manifest as widespread cortical [14] and white matter tract development delays, such as the UF and CC [13].

1.7 Abnormal Brain Development in Neurodevelopmental Disorders for Comorbidities

Although significant and specific alterations in the brains of ASD, ADHD, and DD have been detected, similar clinical symptoms are shared across these neurodevelopmental disorders. Thus, the characterization of shared and unique abnormalities in NDD brains is important for understanding the neural mechanisms underlying comorbidities. Specifically, individuals with ASD often exhibit symptoms of ADHD, i.e., inattention and hyperactivity/impulsivity, and the DSM-V diagnostic criteria permit the coincident diagnosis of ADHD and ASD. Thus, the characterization of shared and unique abnormalities in the brain connectome is important for understanding the neural basis of ASD and any comorbidities. The heightened brain volume observed in autism seems to stem mainly from the unexplained enlargement of white matter, a phenomenon that bears similarities to observations in developmental language disorder. Individuals with ASD often exhibit symptoms of ADHD (e.g., inattention and hyperactivity/impulsivity), and the DSM-V includes a co-diagnosis of ADHD and ASD. A connectomic analysis revealed nodal centrality abnormalities in the precuneus of both ADHD and ASD patients, while other disturbances were disorder-specific, such as an increased connectivity in the right striatum/pallidum and bilateral temporo-limbic areas, respectively. Disorder-specific changes in rich-club organization were also reported: specifically, ASD patients showed hyperconnectivity within and comparable connectivity outside the rich-club structure, whereas ADHD patients showed hypoconnectivity within and hyperconnectivity outside the rich-club structure.

2 Animal Models Translated from Genetic Findings into NDD Patients

A substantial portion of the risk for neurodevelopmental disorders (NDDs) such as autism spectrum disorder (ASD), attention deficit hyperactivity disorder (ADHD), and intellectual disability (ID) is attributed to genetic factors. Emerging genome-wide association studies (GWAS) and large-scale DNA sequencing studies contribute to a better understanding of the genetic basis of NDDs. Animal models are then constructed to transfer clinically identified disease-related genes into in vivo models in the laboratory and replicate features of a human disease phenotype to study the potential relationship between these genes and diseases and their pathogenesis-related mechanisms. Nonetheless, GWAS implicate hundreds of risk loci, contributing to only a limited fraction of the aberrant neurodevelopmental processes. Copy number variations (CNVs, i.e., deletion or duplication) of various susceptible genetic loci are structural variations implicated across many NDDs. In a microarray analysis of 2691 subjects diagnosed with NDDs, 10.5% of cases carried CNVs with potential clinical relevance to NDDs, and ASD had the highest proportion of cases with a clinically relevant CNV (11.4%) [1]. Duplication of chromosome 15q11-q13 (Dup15q) is a common pathogenic CNV for ASD. Jin Nakatani et al. used chromosome engineering to generate a 6.3 Mb duplication of the conserved linkage group on mouse chromosome 7 [2]. This chromosome-engineered mouse model replicated different facets of the human autistic phenotype, such as impaired social interaction and diminished behavioral flexibility [3]. Most patients with Dup15q have maternally derived, and UBE3A, a maternally expressed gene, was considered a driver gene of Dup15q syndrome [4]. However, Necdin was recently identified as a driver gene for developing multiple ASD-like phenotypes in paternal 15q dup mice [5].

Examining the genetics of individuals with NDDs has unveiled a higher occurrence of 16p11.2 CNVs, and both deletions and duplications within the human 16p11.2 gene locus manifest highly pleiotropic phenotypic effects [6, 7]. Mice with 16p11.2 heterozygous null (16p11.2+/−) exhibited repetitive behaviors, hyperactivity, and cognitive deficits [8–10], coinciding with phenotypes present in human 16p11.2 deletion. In addition to mice, some progress has also been made in the transformation from humans to other animals. Among 16p11.2 genes, KCTD13 has been considered a major driver of neurodevelopmental phenotypes [11]. The upregulation of the human transcript in zebrafish embryos pinpointed KCTD13 as the exclusive messenger capable of inducing the microcephaly phenotype linked to the 16p11.2 duplication. Conversely, the downregulation of the same locus contributes to the macrocephalic phenotype associated with the 16p11.2 deletion, mirroring the phenotypes observed in humans [12].

Many genes and mutations are associated with NDDs, indicating a heterogeneous origin of essential neurodevelopmental processes. Advances in gene modifications have enabled the creation of various animal models, which could make the translation from human to animal more readily and better display neural phenotypes. SH3 and multiple ankyrin repeat domains 3 (Shank3) are postsynaptic proteins. Mutation of Shank3 accounts for approximately 1% of idiopathic forms of ASD, and the disruption of SHANK3 is considered to be responsible for autism-like behaviors [13, 14]. Genetic disruption of Shank3 in mice displayed compulsive/repetitive behavior and social deficits, resembling the major features of ASD [15]. CRISPR/Cas9 was applied to generate a Shank3b mutant zebrafish model, which displayed robust autism-like behaviors [16]. Similar behavior phenotypes were observed in macaques with CRISPR-Cas9-mediated generation of SHANK3 mutations [17]. The development of dendritic spines relies on Contactin Associated Protein-like 2 (CNTNAP2), encoded by Cntnap2. Homozygous loss-of-function mutations in CNTNAP2 are significantly linked to ASD. In Cntnap2(-/-) mice, repetitive and restrictive behaviors, along with abnormal social interactions, are observed [18], as well as hyperactivity and epileptic seizures, which are described in many patients with ASD [19].

Moreover, the excitation/inhibition (E/I) imbalance is a common abnormality in animal models with ASD-like behaviors, which is also found in humans. The knockdown of CNTNAP2 leads to reduced excitatory synaptic transmission and impaired social interaction in mice [20], while zebrafish Cntnap2 mutants display GABAergic deficits and significantly greater nighttime activity [21]. A meta-analysis and systematic review suggested that Adhesion G protein-coupled receptor L3 (ADGRL3, also referred to as latrophilin 3 or LPHN3) was linked to an increased risk of ADHD [22]. Samantha L. Regan et al. used CRISPR/Cas9 to create a null mutation of Lphn3 (KO) in Sprague-Dawley rats, which showed persistent hyperactivity and increased acoustic startle [23]. Compared with wild-type controls, Lphn3-/- rats exhibited deficits in impulsive action and working memory [24]. Lphn3.1 regulates locomotor activity and impulsivity in zebrafish, and zebrafish larvae lacking Adgrl3.1 (adgrl3.1-/-) exhibit a pronounced hyperactive phenotype [25, 26]. Besides, Adgrl3-/- mice show increased locomotive activity [27]. The role of Forkhead box p2 (FOXP2) in GABAergic signaling suggests its involvement in neurodevelopmental disorders (NDDs), where various studies highlight disturbances in GABAergic function, disrupting the excitatory/inhibitory (E/I) balance. Additionally, the dopaminergic system is prominently implicated in several NDDs. The dopamine transporter (DAT), situated on the presynaptic membrane in dopaminergic terminals, modulates dopamine (DA) availability. Genetic variations in the DAT gene have been associated with modified behaviors linked to neurodevelopmental disorders (NDDs), including ADHD and ASD [28–30]. Dopamine transporter knockout (DAT-KO) mice are a typical ADHD model, displaying ADHD-related behaviors such as hyperactivity, impulsive-like behavior, and impaired attention [31–34]. A threonine-to-methionine substitution at position 356 (DAT T356M) represents a newly identified mutation associated with ASD in the dopamine transporter (DAT). This mutation is linked to abnormal function, marked by a sustained reverse transport of dopamine. Studies have demonstrated that the T356M variant induces hyperlocomotion in Drosophila melanogaster [29], while the DAT T356M+/+ mice could exhibit typical ADHD and ASD phenotypes such as hyperactivity, repetitive behaviors, and social deficits [35].

2.1 Cross-Species Neuroimaging Reveals Convergent Neural Mechanisms Underlying NDDs

In recent years, progress has been made in imaging techniques, which play a pivotal role in disease detection, therapy monitoring, detection of recurrence, etc. With the availability of ever more powerful techniques (1H-MRS, sMRI, rsfMRI, etc.), new imaging biomarkers may emerge and shape future clinical tools and may also be used in fundamental and translational research to discover mechanisms underlying the diseases.

2.2 Proton Magnetic Resonance Spectroscopy (1H-MRS)

The balance between excitatory and inhibitory neurotransmitter systems is significant to early healthy brain development. Recent studies have proposed E/I imbalance as a possible mechanism underlying neural and behavioral dysfunctions across neurodevelopmental disorders. Glutamate and γ-aminobutyric acid (GABA) are the primary excitatory and inhibitory neurotransmitters of the central nervous system (CNS), respectively. A common noninvasive method to measure glutamate and GABA levels in vivo is proton magnetic resonance spectroscopy (1H-MRS), which can be safely used in both humans and animals and allows direct interspecies comparisons. Certainly, disruptions in glutamate levels within fronto-striatal regions have been corroborated in neurodevelopmental disorders, as evidenced by magnetic resonance spectroscopy (MRS) studies [36].

1H-MRS has revealed alterations in the levels of glutamate and glutamine in the cortex and basal ganglia of children, as well as in the basal ganglia of adults with ASD [37, 38]. Reduced striatal glutamate is recapitulated in several rodent models, such as mice exposed prenatally to valproate and mice and rats carrying mutations in the ASD-associated gene Nlgn3 [39]. Increases in glutamate content in the hippocampus were observed in rats for valproic acid (VPA)- and thalidomide (THAL)-induced models of ASD [40]. Besides, there were significant sexual dimorphisms in the levels of metabolites in the hippocampus and prefrontal cortex [41]. Mice with genetically modified, kinase-inactive Met, serving as a potential model for unraveling neural circuit alterations in ASD, exhibited notable shifts in the balance of glutamate/GABA in the somatosensory thalamocortical system of female Met-Emx1 brains compared to their wild-type counterparts [42].

Compared to the controls, patients with ADHD had reduced mean short interval cortical inhibition (SICI) [43], reduced glutamate concentrations in the prefrontal cortex (PFC) [44], and a significantly lower concentration of glutamate in the basal ganglia [45]. Moreover, glutamate levels in the anterior cingulate cortex (ACC) of individuals with ADHD could potentially be linked to dysfunctions in the striatum during cognitive control [46]. Sensory over-responsivity (SOR), extreme sensitivity to or avoidance of sensory stimuli, is a common feature of neurodevelopmental disorders. The SOR severity correlated negatively with thalamic GABA and positively with somatosensory glutamate [47]. Collectively, these results are consistent with the hypothesis that glutamatergic disturbance plays a significant role in the pathogenesis of NDDs.

2.3 Structural Magnetic Resonance Imaging (sMRI)

Structural magnetic resonance imaging (sMRI) offers information on how anatomical brain development relates to clinical symptoms and functioning in NDDs and can quantify aspects of brain morphometry, such as the size and shape of specific brain structures and regions. A large worldwide collaborative study of brain volume differences between ADHD patients and healthy individuals identified subcortical volumes, including the amygdala, accumbens, hippocampus, and striatum, to be smaller in patients with ADHD [48]. Whole-brain sMRI of controls and patients with ADHD, ASD, and OCD was analyzed in a large-scale study, which revealed smaller intracranial volume (ICV) in children and adolescents with ADHD and thicker frontal cortices in adults with ASD [49]. Brain volume alterations are also well observed in animal models. Moreover, glutamate levels in the anterior cingulate cortex (ACC) of individuals with ADHD could potentially be linked to dysfunctions in the striatum during cognitive control [34]. Spontaneously hypertensive rats (SHRs), a frequently employed animal model for ADHD, tended to have smaller striatal volumes, and the timing of striatal volume changes was found to be approximately equivalent to an age of 7–9 years in children [50]. Moreover, stimulant treatment, highly effective in mitigating ADHD, was identified to possess a causal association with striatal hypertrophy through longitudinal analysis of anatomical MRI in rodents [51].

2.4 Resting-State Functional Magnetic Resonance Imaging (rsfMRI)

Previous studies suggest that NDDs are associated with a large and distributed neural network across multiple brain regions [52]. Impaired sociability is a fundamental symptom in individuals with ASD, potentially originating from atypical connectivity patterns between various brain regions, as suggested by numerous neuroimaging and postmortem investigations. The focus of neuroimaging studies on ASD has progressively shifted toward irregularities in connectivity between brain regions, as opposed to localized functional or structural abnormalities. Anomalies in functional connectivity, assessed through rsfMRI, have consistently been identified in the brains of individuals with ASD.

A comprehensive analysis of intrinsic functional connectivity (iFC) across the entire brain revealed reduced connectivity in corticocortical and interhemispheric functional connectivity. Moreover, it emphasized declines in voxel-wise interhemispheric iFC in subcortical regions, including the thalamus [53]. Despite the availability of animal models that manipulate risk genes at the level of the whole brain, how these genes affect specific neural circuits involved in NDDs needs further exploration. A mutation specifically targeted to a circuit (ctMUT) associated with the ASD-risk gene Shank3 within a unidirectional projection from the prefrontal cortex to the basolateral amygdala induces changes in spine morphology and disrupts the excitatory/inhibitory (E/I) balance of the circuit. A higher mPFC-amygdala functional connectivity was discovered in individuals with ASD and Shank3 ctMUT mice, which is associated with reduced sociability [54]. Elyza Kelly et al. delineated a circuit from cerebellar cortical areas Right crus 1 (Rcrus1) and posterior vermis through the cerebellar nuclei and ventromedial thalamus and culminating in the mPFC, whose modulation induced social deficits and repetitive behaviors [55]. In addition, rsfMRI showed a significantly reduced primary somatosensory cortex (S1)-hippocampal connectivity in Shank3b-/- adult mice, which was associated with exhibiting atypical reactions to touch, a characteristic aspect of ASD. The analysis of resting-state functional magnetic resonance imaging (rsfMRI) in Chd8 heterozygous mutant mice, displaying myelination defects and behaviors akin to ASD, demonstrated alterations in functional connectivity within the cortex, hippocampus, and amygdala, with the magnitude of change correlating with alterations in social interaction [56]. Moreover, Cntnap2(-/-) mice display diminished functional connectivity both locally and over long distances in the prefrontal and midline brain regions. Significantly, viral tracing experiments unveiled a decreased occurrence of neural clusters projecting to the prefrontal cortex in the cingulate cortex of Cntnap2-/- mutants [57].

As a cross-species technique, neuroimaging can be performed repeatedly within subjects and used to translate findings from the preclinical to clinical laboratory using the same imaging biomarker. In parallel, findings from the clinic can also be translated back to the preclinical laboratory to develop animal models, test whether a compound is able to modulate brain metabolite abnormalities associated with NDDs, and help uncover the pathology and anatomy of diseases.

2.5 Cerebral Organoid: An Ideal Approach for Modeling NDDs

Addressing neurodevelopmental disorders (NDDs) is notably challenging due to the intricate pathophysiology, primarily arising from multigenic causation, making in vitro modeling of the conditions arduous. The scarcity of subjects for in-depth analysis and constraints on disease-specific animal models impede comprehensive insights into NDDs. Nonetheless, the introduction of patient-derived induced pluripotent stem cells (iPSCs) presents a valuable avenue for comprehending the molecular progression of NDDs across cellular, tissue, and organ levels [58, 59]. Chun et al. used three-dimensional (3D) genomic data and generated high-resolution maps of promoter interactions in human iPSC-derived neural progenitor cells (NPCs) and neurons, connecting neurodevelopmental disorder-associated signals to putative causal effector genes [60]. To overcome the limitations of animal models concerning genetic heterogeneity with humans, human iPSC-derived 3D brain organoids have been developed [61]. The combination of available tools, including CRISPR-Cas9, cell-based two-dimensional (2D) models, and 3D brain organoid models, makes the generation of cell lines with patient-specific mutations or the "repair" of patient-specific mutations to generate an isogenic control cell line possible, which may contribute to decoding the pathophysiology of NDDs [62, 63]. Madeline A.L. et al. utilized patient-derived iPSCs and shRNA in cerebral organoids to model CDK5RAP2-dependent pathogenesis of microcephaly, a disorder that has been difficult to model in mice [64]. Yuki Miura and colleagues devised a technique for producing three-dimensional spheroids that mimic distinct nervous system domains, facilitating their physical integration to support axonal projections and the formation of synaptic connections. This method will prove valuable for investigating the assembly of neural circuits and for modeling Neurodevelopmental Disorders (NDDs) using induced pluripotent stem cells (iPSCs) [65]. In a study of RAB39b-associated macrocephaly/autism, cerebral organoids were created, and a dysregulated PI3K-AKT-mTOR signaling that leads to a dysregulation of cortical neurogenesis was discovered, contributing to macrocephaly and autistic-like behaviors and providing new insights into the mechanisms of neurodevelopmental dysregulation and common pathways across species [66]. Besides, the cerebral organoids follow the in vivo timeline, which suggests that they can simulate early (8–10 gestation week) to mid-fetal (22–24/35 gestation week) human brain development as well as different region-specific organoids such as the forebrain, midbrain, cortex, and hypothalamus through growth factors and morphogens [67].

So far, there have been a variety of NDDs modeled by cerebral organoids. The chromodomain helicase DNA-binding protein 8 (CHD8) has consistently emerged as a candidate gene for Autism Spectrum Disorder (ASD). Consequently, a CHD8+/− induced pluripotent stem cell (iPSC) line was created using CRISPR/Cas9 technology, and this line was employed to generate cerebral organoids for ASD modeling [68]. While WDR62 mutant cerebral organoids model human microcephaly, KATNB1 mutant cerebral organoids model human lissencephaly, and PTEN mutant cerebral organoids model human macrocephaly [69, 70].

Daviaud N et al. achieved successful transplantation of cerebral organoids into the mouse cortex, demonstrating improved survival and robust vascularization compared to in vitro models based solely on induced pluripotent stem cells (iPSCs). This presents a distinctive experimental approach for modeling Neurodevelopmental Disorders (NDDs) within the context of vascularized cortical tissue [71]. Hartlaub AM et al. transplanted human brain organoids into adult mouse brains, where they demonstrated integration, viability, long-term survival, vascularization, functional neuronal activity, and synaptic connectivity between the grafted organoid and the host, providing numerous avenues for future research [72]. Trujillo CA et al. conjecture that patient-derived cerebral organoids may be used to diagnose or test various drugs and doses in the future [73]. Nevertheless, well in advance of that, we must understand how to extrapolate findings from cerebral organoids in the laboratory to the complexities of the human brain.

Furthermore, the concept of 3D bioprinting has been suggested to bring about a transformation in the realm of neural tissue engineering, enhancing the efficiency and consistency in the production of cerebral organoids [74], but there are still many obstacles to cross to realize this immense potential. The latest progress in cerebral organoids has paved the way for the creation of innovative, more physiologically accurate human models. These preclinical models are crucial in facilitating the translation of fundamental research into new therapeutic approaches for individuals with neurodevelopmental disorders (NDDs). Moreover, the advent of patient-derived cerebral organoids provides the opportunity for personalized treatment, allowing the assessment of drug responses on these organoids in a preclinical laboratory setting.

2.6 Therapeutic Translation of NDDs from Animal to Human Studies

Following the initial identification of the first animal model inducing autistic-like behavior with valproic acid (VPA), numerous genetic and nongenetic experimental animal models for autism spectrum disorder (ASD) and other neurodevelopmental disorders (NDDs) have been uncovered. The subsequent challenge is to bridge the gap between biomedical discoveries made on various models (such as animals and cerebral organoids) and the development of effective clinical treatments for NDDs. Currently, there are both pharmacological and nonpharmacological interventions for NDDs, offering partial relief from core symptoms or management of comorbid conditions associated with these disorders [75].

Animal models continue to be the primary means of assessing the efficacy of new therapeutic drugs prior to human clinical trials. However, the translation of successful outcomes from animal models to humans poses a significant challenge, and the incorporation of these findings into clinical therapies has yielded inconsistent results. Social interaction and communication are partly modulated by neuropeptides, such as vasopressin and oxytocin. Targeting the vasopressin 1a (V1a) receptor in a rat model of ASD has demonstrated the potential to enhance social interaction [76]. Balovaptan, an orally administered, brain-penetrant, selective V1a receptor antagonist, demonstrated dose-dependent and clinically meaningful improvements in the Vasopressin Antagonist to Improve Social Communication in Autism (VANILLA) study compared to the placebo [77]. The neuropeptide oxytocin also increases social approach and social memory in animals, both of which are impaired in ASD patients. Oxytocin administration rescued social deficits in the Cntnap2 knockout (KO) mouse model [78], accompanied by intensely activated regions of the brain and restored normalized connectivity patterns [79]. Multiple clinical trials have demonstrated the capacity of oxytocin to augment social reinforcement learning and ameliorate repetitive behaviors and social functioning [80–83]. These results have been supported by functional neuroimaging studies involving ASD patients, which have shown altered regional brain activation after the administration of intranasal oxytocin [84, 85].

Cannabidiol (CBD) interacts with the endogenous cannabinoid (eCB) system, which modulates excitatory glutamate and inhibitory GABA transmission, both of which are significant treatment targets in NDDs [86]. There is evidence indicating that CBD successfully mitigated social deficits resembling autism in a thoroughly validated mouse genetic model of Dravet syndrome (DS) [87, 88]. Clinical trials have recently investigated CBD therapy for ASD, revealing a reduction in GABA levels in prefrontal and subcortical regions in patients with ASD [89]. This was accompanied by extensive alterations in functional connectivity with various subcortical and cortical targets [90]. Barchel et al. reported improvements in ASD comorbid symptoms such as hyperactivity and anxiety in oral CBD treatment trials with ASD children [91]. Clinically positive results have also been obtained in children and adolescents with fragile X syndrome [92].

Distinct origins may contribute to the behavioral abnormalities of NDDs, and it is noteworthy that the gut microbiome can exert influence on specific behaviors. Buffington SA et al. identified that the hyperactivity phenotype in Cntnap2-/- mice is influenced by host genetics, whereas the social-behavior phenotype is mediated by the gut microbiome [93]. Transplanting gut microbiota from individuals with and without ADHD into mice revealed a reduction in the structural integrity of both white and gray matter regions, including the internal capsule and hippocampus. Additionally, mice colonized with ADHD microbiota showed decreased rsfMRI-based connectivity between the right motor and right visual cortices [94]. Changes in the gut microbiota are implicated in the cognitive and social dysfunctions linked to maternal obesity, a factor associated with neurodevelopmental disorders (NDDs) in offspring. Interventions involving dietary fiber for either dams or 3-week-old offspring have shown promise in reversing abnormal neurocognitive and social behaviors in the offspring, offering a potential therapeutic avenue for neurodevelopmental disorders [95]. Altered gut microbial composition is common in ASD individuals [96]. Consumption of a maternal high-fat diet (MHFD) has been shown to induce social deficits in mice and alter the gut microbiota of offspring. This alteration is marked by a decrease in the presence of the commensal bacterial species *L. reuteri* [97, 98]. Hence, probiotic intervention has the potential to alleviate particular behavioral abnormalities linked to neurodevelopmental disorders (NDDs). Correspondingly, *L. reuteri* has demonstrated the ability to ameliorate social deficits in diverse mouse models of ASD [99], which holds promise for novel therapies in patients. Compared with the placebo group, Lactobacillus plantarum PS128 (PS128) could ameliorate some ASD symptoms, primarily those associated with disruptive and rule-breaking behaviors and hyperactivity/impulsivity [100]. An orally administered gastrointestinal-restricted adsorbent (AB-2004), exhibiting an affinity for small aromatic or phenolic molecules, was shown to mitigate anxiety-like behaviors in mice induced by a gut microbial metabolite [101].

Nevertheless, preclinical animal research has limitations, making it challenging to accurately anticipate clinical trial success. Despite achievements in animal studies, around 85% of initial clinical trials for new drugs end in failure. Even among those that progress to Phase III trials, nearly 50% ultimately fail due to an inability to demonstrate treatment efficacy [102]. While certain individuals continue to grapple with finding avenues to translate success, others are prompted to explore alternative approaches. Neurofeedback relies on a brain-computer interface (BCI) and is executed through a software system and processing pipeline comprising five elements: data acquisition, online data processing, online feature extraction, online feedback generation, and learning [103]. Neurofeedback involves assessing the participant's brain activity, preprocessing it, and converting it into signals that are presented to the participant. This feedback allows the participant to acquire the skill of self-regulating their brain activity, directly influencing the neural mechanisms of cognition and behavior. Recognized as a promising nonpharmacological and noninvasive intervention, neurofeedback holds potential for the long-term management of ADHD symptoms [104]. Given its capacity to enhance core symptoms like inattention, hyperactivity, impulsivity, and learning difficulties, while promoting self-regulation of cognition, emotion, and behavior, neurofeedback is recognized for its potential [105, 106]. Importantly, it achieves these benefits without the occurrence of significant adverse events related to treatment or trials [107].

Watanabe M. et al. applied the organoid model to study microcephaly resulting from ZIKV infection, testing three drug candidates, and found something different from previous studies [108]. While the predictive validity of cerebral organoids requires additional confirmation, they offer a potential platform for validating the outcomes of compound screening in the pursuit of drug candidates. Discrepancies in modeling among animal models, cerebral organoids, and humans may account for the challenges in translating controlled translational findings directly into a clinical context. However, there is no need to lose confidence in the face of these challenges. Various strategies have been suggested to enhance the rigor of preclinical studies and increase the success rate of transitioning to clinical applications. It is anticipated that future efforts will yield more success in this regard.

References

1. Zarrei M, et al. A large data resource of genomic copy number variation across neurodevelopmental disorders. NPJ Genom Med. 2019;4:26.
2. Nakatani J, et al. Abnormal behavior in a chromosome-engineered mouse model for human 15q11-13 duplication seen in autism. Cell. 2009;137(7):1235–46.
3. Nakai N, et al. Serotonin rebalances cortical tuning and behavior linked to autism symptoms in 15q11-13 CNV mice. Sci Adv. 2017;3(6):e1603001.
4. Baron CA, et al. Genomic and functional profiling of duplicated chromosome 15 cell lines reveal regulatory alterations in UBE3A-associated ubiquitin-proteasome pathway processes. Hum Mol Genet. 2006;15(6):853–69.
5. Tamada K, et al. Genetic dissection identifies Necdin as a driver gene in a mouse model of paternal 15q duplications. Nat Commun. 2021;12(1):4056.
6. Green Snyder L, et al. Autism spectrum disorder, developmental and psychiatric features in 16p11.2 duplication. J Autism Dev Disord. 2016;46(8) 2734–48.
7. D'Angelo D, et al. Defining the effect of the 16p11.2 duplication on cognition, behavior, and medical comorbidities. JAMA Psychiatry. 2016;73(1):20–30.
8. Brunner D, et al. Comprehensive analysis of the 16p11.2 deletion and null Cntnap2 mouse models of autism spectrum disorder. PLoS One. 2015;10(8):e0134572.
9. Arbogast T, et al. Reciprocal effects on neurocognitive and metabolic phenotypes in mouse models of 16p11.2 deletion and duplication syndromes. PLoS Genet. 2016;12(2):e1005709.
10. Wang W, et al. Chemogenetic activation of prefrontal cortex rescues synaptic and behavioral deficits in a mouse model of 16p11.2 deletion syndrome. J Neurosci. 2018;38(26):5939–48.
11. Escamilla CO, et al. Kctd13 deletion reduces synaptic transmission via increased RhoA. Nature. 2017;551(7679):227–31.
12. Golzio C, et al. KCTD13 is a major driver of mirrored neuroanatomical phenotypes of the 16p11.2 copy number variant. Nature. 2012;485(7398):363–7.
13. Sanders SJ, et al. Insights into autism spectrum disorder genomic architecture and biology from 71 risk loci. Neuron. 2015;87(6):1215–33.
14. Moessner R, et al. Contribution of SHANK3 mutations to autism spectrum disorder. Am J Hum Genet. 2007;81(6):1289–97.
15. Peca J, et al. Shank3 mutant mice display autistic-like behaviours and striatal dysfunction. Nature. 2011;472(7344):437–42.
16. Liu CX, et al. CRISPR/Cas9-induced shank3b mutant zebrafish display autism-like behaviors. Mol Autism. 2018;9:23.
17. Zhou Y, et al. Atypical behaviour and connectivity in SHANK3-mutant macaques. Nature. 2019;570(7761):326–31.
18. Penagarikano O, et al. Absence of CNTNAP2 leads to epilepsy, neuronal migration abnormalities, and core autism-related deficits. Cell. 2011;147(1):235–46.
19. Frye RE, et al. Neuropathological mechanisms of seizures in autism spectrum disorder. Front Neurosci. 2016;10:192.
20. Sacai H, et al. Autism spectrum disorder-like behavior caused by reduced excitatory synaptic transmission in pyramidal neurons of mouse prefrontal cortex. Nat Commun. 2020;11(1):5140.
21. Hoffman EJ, et al. Estrogens suppress a behavioral phenotype in zebrafish mutants of the autism risk gene, CNTNAP2. Neuron. 2016;89(4):725–33.
22. Bruxel EM, et al. Meta-analysis and systematic review of ADGRL3 (LPHN3) polymorphisms in ADHD susceptibility. Mol Psychiatry. 2021;26(6):2277–85.
23. Regan SL, et al. Knockout of latrophilin-3 in Sprague-Dawley rats causes hyperactivity, hyperreactivity, under-response to amphetamine, and disrupted dopamine markers. Neurobiol Dis. 2019;130:104494.
24. Sable HJK, et al. An assessment of executive function in two different rat models of attention-deficit hyperactivity disorder: spontaneously hypertensive versus Lphn3 knockout rats. Genes Brain Behav. 2021;20(8):e12767.
25. Sveinsdottir HS, et al. Novel non-stimulants rescue hyperactive phenotype in an adgrl3.1 mutant zebrafish model of ADHD. Neuropsychopharmacology. 2022;48(8):1155–63.
26. Lange M, et al. The ADHD-linked gene Lphn3.1 controls locomotor activity and impulsivity in zebrafish. Mol Psychiatry. 2012;17(9):855.

27. Mortimer N, et al. Dissociation of impulsivity and aggression in mice deficient for the ADHD risk gene Adgrl3: evidence for dopamine transporter dysregulation. Neuropharmacology. 2019;156:107557.
28. Cook EH Jr, et al. Association of attention-deficit disorder and the dopamine transporter gene. Am J Hum Genet. 1995;56(4):993–8.
29. Hamilton PJ, et al. De novo mutation in the dopamine transporter gene associates dopamine dysfunction with autism spectrum disorder. Mol Psychiatry. 2013;18(12):1315–23.
30. Faraone SV, et al. Molecular genetics of attention-deficit/hyperactivity disorder. Biol Psychiatry. 2005;57(11):1313–23.
31. Gainetdinov RR, Caron MG. Genetics of childhood disorders: XXIV. ADHD, part 8: hyperdopaminergic mice as an animal model of ADHD. J Am Acad Child Adolesc Psychiatry. 2001;40(3):380–2.
32. Regan SL, Williams MT, Vorhees CV. Review of rodent models of attention deficit hyperactivity disorder. Neurosci Biobehav Rev. 2022;132:621–37.
33. Weiss S, et al. Nicotine improves cognitive deficits of dopamine transporter knockout mice without long-term tolerance. Neuropsychopharmacology. 2007;32(12):2465–78.
34. Reinwald JR, et al. Dopamine transporter silencing in the rat: systems-level alterations in striato-cerebellar and prefrontal-midbrain circuits. Mol Psychiatry. 2022;27(4):2329–39.
35. DiCarlo GE, et al. Autism-linked dopamine transporter mutation alters striatal dopamine neurotransmission and dopamine-dependent behaviors. J Clin Invest. 2019;129(8):3407–19.
36. Naaijen J, et al. Fronto-striatal glutamatergic compounds in compulsive and impulsive syndromes: a review of magnetic resonance spectroscopy studies. Neurosci Biobehav Rev. 2015;52:74–88.
37. Bejjani A, et al. Elevated glutamatergic compounds in pregenual anterior cingulate in pediatric autism spectrum disorder demonstrated by 1H MRS and 1H MRSI. PLoS One. 2012;7(7):e38786.
38. Brown MS, et al. Increased glutamate concentration in the auditory cortex of persons with autism and first-degree relatives: a (1)H-MRS study. Autism Res. 2013;6(1):1–10.
39. Horder J, et al. Glutamate and GABA in autism spectrum disorder-a translational magnetic resonance spectroscopy study in man and rodent models. Transl Psychiatry. 2018;8(1):106.
40. Zieminska E, et al. Glutamate, glutamine and GABA levels in rat brain measured using MRS, HPLC and NMR methods in study of two models of autism. Front Mol Neurosci. 2018;11:418.
41. Ferreira H, et al. Sex-dependent social and repetitive behavior and neurochemical profile in mouse model of autism spectrum disorder. Meta. 2022;12(1):71.
42. Tang S, et al. Altered forebrain functional connectivity and neurotransmission in a kinase-inactive met mouse model of autism. Mol Imaging. 2019;18:1536012118821034.
43. Gilbert DL, et al. Motor cortex inhibition: a marker of ADHD behavior and motor development in children. Neurology. 2011;76(7):615–21.
44. Dramsdahl M, et al. Adults with attention-deficit/hyperactivity disorder – a brain magnetic resonance spectroscopy study. Front Psych. 2011;2:65.
45. Maltezos S, et al. Glutamate/glutamine and neuronal integrity in adults with ADHD: a proton MRS study. Transl Psychiatry. 2014;4(3):e373.
46. Naaijen J, et al. Anterior cingulate cortex glutamate and its association with striatal functioning during cognitive control. Eur Neuropsychopharmacol. 2018;28(3):381–91.
47. Wood ET, et al. Sensory over-responsivity is related to GABAergic inhibition in thalamocortical circuits. Transl Psychiatry. 2021;11(1):39.
48. Hoogman M, et al. Subcortical brain volume differences in participants with attention deficit hyperactivity disorder in children and adults: a cross-sectional mega-analysis. Lancet Psychiatry. 2017;4(4):310–9.
49. Boedhoe PSW, et al. Subcortical brain volume, regional cortical thickness, and cortical surface area across disorders: findings from the ENIGMA ADHD, ASD, and OCD working groups. Am J Psychiatry. 2020;177(9):834–43.
50. Hsu JW, et al. Striatal volume changes in a rat model of childhood attention-deficit/hyperactivity disorder. Psychiatry Res. 2010;179(3):338–41.
51. Biezonski D, et al. Longitudinal magnetic resonance imaging reveals striatal hypertrophy in a rat model of long-term stimulant treatment. Transl Psychiatry. 2016;6(9):e884.
52. Kast RJ, Levitt P. Precision in the development of neocortical architecture: from progenitors to cortical networks. Prog Neurobiol. 2019;175:77–95.
53. Di Martino A, et al. The autism brain imaging data exchange: towards a large-scale evaluation of the intrinsic brain architecture in autism. Mol Psychiatry. 2014;19(6):659–67.
54. Kim S, et al. Neural circuit pathology driven by Shank3 mutation disrupts social behaviors. Cell Rep. 2022;39(10):110906.
55. Kelly E, et al. Regulation of autism-relevant behaviors by cerebellar-prefrontal cortical circuits. Nat Neurosci. 2020;23(9):1102–10.
56. Kawamura A, et al. Chd8 mutation in oligodendrocytes alters microstructure and functional connectivity in the mouse brain. Mol Brain. 2020;13(1):160.
57. Liska A, et al. Homozygous loss of autism-risk gene CNTNAP2 results in reduced local and long-range prefrontal functional connectivity. Cereb Cortex. 2018;28(4):1141–53.
58. Sabitha KR, et al. Patient-derived iPSC modeling of rare neurodevelopmental disorders: molecular pathophysiology and prospective therapies. Neurosci Biobehav Rev. 2021;121:201–19.
59. Hohmann SS, Ilieva M, Michel TM. In vitro models for ASD-patient-derived iPSCs and cerebral organoids. Prog Mol Biol Transl Sci. 2020;173:355–75.

60. Su C, et al. 3D promoter architecture re-organization during iPSC-derived neuronal cell differentiation implicates target genes for neurodevelopmental disorders. Prog Neurobiol. 2021;201:102000.
61. Del Dosso A, et al. Upgrading the physiological relevance of human brain organoids. Neuron. 2020;107(6):1014–28.
62. Whiteley JT, et al. Reaching into the toolbox: stem cell models to study neuropsychiatric disorders. Stem Cell Reports. 2022;17(2):187–210.
63. Fischer J, et al. Genetic modification of brain organoids. Front Cell Neurosci. 2019;13:558.
64. Lancaster MA, et al. Cerebral organoids model human brain development and microcephaly. Nature. 2013;501(7467):373–9.
65. Miura Y, et al. Engineering brain assembloids to interrogate human neural circuits. Nat Protoc. 2022;17(1):15–35.
66. Zhang W, et al. Cerebral organoid and mouse models reveal a RAB39b-PI3K-mTOR pathway-dependent dysregulation of cortical development leading to macrocephaly/autism phenotypes. Genes Dev. 2020;34(7–8):580–97.
67. Forsberg SL, et al. Epigenetics and cerebral organoids: promising directions in autism spectrum disorders. Transl Psychiatry. 2018;8(1):14.
68. Wang P, et al. CRISPR/Cas9-mediated heterozygous knockout of the autism gene CHD8 and characterization of its transcriptional networks in cerebral organoids derived from iPS cells. Mol Autism. 2017; 8:11.
69. Zhang W, et al. Modeling microcephaly with cerebral organoids reveals a WDR62-CEP170-KIF2A pathway promoting cilium disassembly in neural progenitors. Nat Commun. 2019;10(1):2612.
70. Li Y, et al. Induction of expansion and folding in human cerebral organoids. Cell Stem Cell. 2017;20:385–396.e3.
71. Daviaud N, et al. Vascularization and engraftment of transplanted human cerebral organoids in mouse cortex. eNeuro. 2018;5(6):ENEURO.0219-18.2018.
72. Hartlaub AM, McElroy CA, Maitre NL, Hester ME. Modeling human brain circuitry using pluripotent stem cell platforms. Front Pediatr. 2019;7:57.
73. Trujillo CA, Muotri AR. Brain organoids and the study of neurodevelopment. Trends Mol Med. 2018;24(12):982–90.
74. De la Vega L, et al. 3D bioprinting models of neural tissues: the current state of the field and future directions. Brain Res Bull. 2019;150:240–9.
75. Sharma SR, et al. Autism spectrum disorder: classification, diagnosis and therapy. Pharmacol Ther. 2018;190:91–104.
76. Felix-Ortiz AC, Febo M. Gestational valproate alters BOLD activation in response to complex social and primary sensory stimuli. PLoS One. 2012;7(5):e37313.
77. Bolognani F, et al. A phase 2 clinical trial of a vasopressin V1a receptor antagonist shows improved adaptive behaviors in men with autism spectrum disorder. Sci Transl Med. 2019;11(491):eaat7838.
78. Penagarikano O, et al. Exogenous and evoked oxytocin restores social behavior in the Cntnap2 mouse model of autism. Sci Transl Med. 2015;7(271):271ra8.
79. Choe KY, et al. Oxytocin normalizes altered circuit connectivity for social rescue of the Cntnap2 knockout mouse. Neuron. 2022;110(5):795–808 e6.
80. Kruppa JA, et al. Neural modulation of social reinforcement learning by intranasal oxytocin in male adults with high-functioning autism spectrum disorder: a randomized trial. Neuropsychopharmacology. 2019;44(4):749–56.
81. Korisky A, Gordon I, Goldstein A. Oxytocin impacts top-down and bottom-up social perception in adolescents with ASD: a MEG study of neural connectivity. Mol Autism. 2022;13(1):36.
82. Bernaerts S, et al. Behavioral effects of multiple-dose oxytocin treatment in autism: a randomized, placebo-controlled trial with long-term follow-up. Mol Autism. 2020;11(1):6.
83. Yamasue H, et al. Effect of intranasal oxytocin on the core social symptoms of autism spectrum disorder: a randomized clinical trial. Mol Psychiatry. 2020;25(8):1849–58.
84. Gordon I, et al. Oxytocin enhances brain function in children with autism. Proc Natl Acad Sci USA. 2013;110(52):20953–8.
85. Greene RK, et al. The effects of intranasal oxytocin on reward circuitry responses in children with autism spectrum disorder. J Neurodev Disord. 2018;10(1):12.
86. Kwan Cheung KA, Mitchell MD, Heussler HS. Cannabidiol and neurodevelopmental disorders in children. Front Psych. 2021;12:643442.
87. Kaplan JS, et al. Cannabidiol attenuates seizures and social deficits in a mouse model of Dravet syndrome. Proc Natl Acad Sci USA. 2017;114(42):11229–34.
88. Wirrell EC, Nabbout R. Recent advances in the drug treatment of Dravet syndrome. CNS Drugs. 2019;33(9):867–81.
89. Pretzsch CM, et al. Effects of cannabidiol on brain excitation and inhibition systems; a randomised placebo-controlled single dose trial during magnetic resonance spectroscopy in adults with and without autism spectrum disorder. Neuropsychopharmacology. 2019a;44(8):1398–405.
90. Pretzsch CM, et al. The effect of cannabidiol (CBD) on low-frequency activity and functional connectivity in the brain of adults with and without autism spectrum disorder (ASD). J Psychopharmacol. 2019b;33(9):1141–8.
91. Barchel D, et al. Oral cannabidiol use in children with autism spectrum disorder to treat related symptoms and co-morbidities. Front Pharmacol. 2018;9:1521.

92. Heussler H, et al. A phase 1/2, open-label assessment of the safety, tolerability, and efficacy of transdermal cannabidiol (ZYN002) for the treatment of pediatric fragile X syndrome. J Neurodev Disord. 2019;11(1):16.
93. Buffington SA, et al. Dissecting the contribution of host genetics and the microbiome in complex behaviors. Cell. 2021;184(7):1740–1756 e16.
94. Tengeler AC, et al. Gut microbiota from persons with attention-deficit/hyperactivity disorder affects the brain in mice. Microbiome. 2020;8(1):44.
95. Liu X, et al. High-fiber diet mitigates maternal obesity-induced cognitive and social dysfunction in the offspring via gut-brain axis. Cell Metab. 2021;33(5):923–938 e6.
96. Vuong HE, Hsiao EY. Emerging roles for the gut microbiome in autism spectrum disorder. Biol Psychiatry. 2017;81(5):411–23.
97. Buffington SA, et al. Microbial reconstitution reverses maternal diet-induced social and synaptic deficits in offspring. Cell. 2016;165(7):1762–75.
98. Sun J, et al. High-fat-diet-induced obesity is associated with decreased antiinflammatory lactobacillus reuteri sensitive to oxidative stress in mouse Peyer's patches. Nutrition. 2016;32(2):265–72.
99. Sgritta M, et al. Mechanisms underlying microbial-mediated changes in social behavior in mouse models of autism spectrum disorder. Neuron. 2019;101(2):246–259 e6.
100. Liu YW, et al. Effects of lactobacillus plantarum PS128 on children with autism spectrum disorder in Taiwan: a randomized, double-blind, placebo-controlled trial. Nutrients. 2019;11(4):820.
101. Stewart Campbell A, et al. Safety and target engagement of an oral small-molecule sequestrant in adolescents with autism spectrum disorder: an open-label phase 1b/2a trial. Nat Med. 2022;28(3):528–34.
102. Beck AP, Meyerholz DK. Evolving challenges to model human diseases for translational research. Cell Tissue Res. 2020;380(2):305–11.
103. Enriquez-Geppert S, et al. Neurofeedback as a treatment intervention in ADHD: current evidence and practice. Curr Psychiatry Rep. 2019;21(6):46.
104. Lambez B, et al. Non-pharmacological interventions for cognitive difficulties in ADHD: a systematic review and meta-analysis. J Psychiatr Res. 2020;120:40–55.
105. Roy S, et al. Effectiveness of neurofeedback training, behaviour management including attention enhancement training and medication in children with attention-deficit/hyperactivity disorder – a comparative follow up study. Asian J Psychiatr. 2022;76:103133.
106. Heinrich H, et al. Effects of neurofeedback on the dysregulation profile in children with ADHD: SCP NF meets SDQ-DP – a retrospective analysis. Psychol Med. 2020;50(2):258–63.
107. Schonenberg M, et al. Neurofeedback, sham neurofeedback, and cognitive-behavioural group therapy in adults with attention-deficit hyperactivity disorder: a triple-blind, randomised, controlled trial. Lancet Psychiatry. 2017;4(9):673–84.
108. Watanabe M, et al. Self-organized cerebral organoids with human-specific features predict effective drugs to combat Zika virus infection. Cell Rep. 2017;21(2):517–32.

Cellular and Molecular Biology of Neurodevelopment

Mohammad Zonuzirad

Abstract

The development of the brain occurs through a meticulously coordinated series of stages. By the time of birth, neurogenesis and the establishment of the overall brain structure are largely completed. However, the growth of glial cells, synaptogenesis, synapse pruning, and myelination take place in the postnatal period. Research now shows that neurogenesis happens before gliogenesis, indicating a well-timed mechanism governed by intricate interactions between intrinsic factors (such as gene epigenetic modifications) and extrinsic influences (contact-mediated elements). The precision and effectiveness of synaptic connections and neural circuits in the central nervous system stand out as one of its most notable characteristics, resulting from the development and maturation of synapses to facilitate overall brain function, complex cognitive processes, and behaviors.

During the initial 8 years of life, the brain undergoes significant development and alterations, setting the stage for future learning, well-being, and overall success. Both positive and negative encounters can accumulate to influence a child's growth trajectory, with lasting impacts. To foster the physical and cognitive development of their offspring, parents and caregivers require adequate support and appropriate resources. Providing optimal care for children, commencing prior to birth and extending throughout childhood, is crucial for ensuring optimal brain development and the realization of its full potential. Prolonged exposure to stress and trauma may result in enduring adverse effects on the child's brain, whereas engaging in activities such as talking, reading, and playing can promote brain growth.

The study of normal human brain development and the identification of abnormalities that lead to disorders are hindered by the challenges associated with studying the brain at the molecular and cellular level. Currently, our understanding of brain development in humans is primarily based on a limited number of postmortem studies and a larger number of in vivo brain imaging studies. To gain insights into the molecular and cellular development of the human brain, researchers heavily rely on studies conducted on animal models, which are then extrapolated to human developmental timelines.

From a visible anatomical perspective, the human brain undergoes a complex developmental process during gestation, resulting in the formation of distinct anatomical regions and a highly convoluted surface. Similarly, at the cellular level, the human brain is a

M. Zonuzirad (✉)
Pediatric Department, Aram Hospital, Karaj, Iran

© The Author(s), under exclusive license to Springer Nature Singapore Pte Ltd. 2024
R. Kelishadi (ed.), *Nature, Nurture, and Neurodevelopment*,
https://doi.org/10.1007/978-981-97-7433-3_3

sophisticated and layered structure composed of various types of cells that possess specific interconnections. The growth and guidance of axons, for instance, are influenced by a combination of signals that promote or inhibit growth, allowing them to navigate and connect with the appropriate target cells. Additionally, the survival of neurons is partially dependent on the availability of neurotrophic factors, which are molecules derived from the target cells themselves.

During certain developmental stages, environmental factors can have a significant impact on specific developmental processes. These periods, known as "sensitive" or "critical" periods, are crucial for proper brain development and can also make the developing brain more susceptible to negative influences from the environment. In humans, sensitive periods are most commonly associated with disruptions in the development of basic sensory and motor functions. The consequences of environmental deprivation and lack of human interaction may be particularly severe during the early stages of development.

Keywords

Cellular biology · Molecular biology · Neurodevelopment · Neuroplasticity · Neuronal plasticity · Neurogenesis · Learning · Brain rewiring · Brain development · Neural circuit · Synaptic transmission · Neurotransmitter switching · Use it or lose it · BDNF · Neurotrophic factor · Nerve growth factor · Signaling · Radial glial cells

Neuroplasticity, also referred to as plasticity, denotes the ability of the nervous system to adjust and change in reaction to outside stimuli [1]. Plastic changes encompass modifications to neuronal networks, adjustments in neuronal and synaptic connectivity, production of fresh neurons, and additional neurobiochemical alterations that transpire in reaction to our encounters [2].

The Polish neuroscientist Jerzy Konorski coined the term "neuroplasticity" in 1948 to elucidate the discernible alterations in brain structure. However, it was not until the 1960s that this term gained extensive recognition and utilization. Despite Santiago Ramón y Cajal, renowned as the "father of neuroscience," contemplating "neuronal plasticity" in the early 1900s [3], the concept predates his time [4]. Contrary to what was widely believed at the time, he noticed that a person's brain might change as they grew older.

During the 1960s, a significant finding emerged indicating that neurons undergo a process of "reorganization" in response to stressful events. Further investigations have substantiated that stress has the potential to alter both the physical composition and functional dynamics of the brain [3].

Stress might actually lead to the death of brain cells, according to research from the late 1990s, but this conclusion is still not completely definitive. For a while, people believed that the brain was an "irreplaceable organ," meaning that humans could only have a limited number of brain cells. As we grew older, these cells would gradually diminish, regardless of our efforts to preserve them. According to Ramón y Cajal, the pathways in the adult brain were considered permanent, complete, and unalterable. It seemed that everything could perish, but nothing could be regenerated [3].

Research indicates that there exist additional ways in which brain cells can deteriorate, adjust, interconnect, and possibly even renew or restore. This process is termed "neurogenesis."

1 Neuroplasticity Versus Neurogenesis

Neurogenesis and neuroplasticity, although distinct concepts, are intricately intertwined. Neuroplasticity entails the brain's capacity to forge fresh connections and pathways, thereby altering the configuration of its circuits. Conversely, neurogenesis is an even more remarkable attribute of the brain, enabling it to generate novel neurons [5].

The concept of neurogenesis presents itself as a more attractive option for various medical applications. While utilizing existing resources is valuable, the ability to effectively regenerate neurons lost due to various conditions could potentially revolutionize dementia treatment, traumatic brain injury rehabilitation, and other unexplored areas of medical research.

1.1 Neuroplasticity and Learning

Learning and neuroplasticity are inherently intertwined due to the brain's ability to form new neural connections. As we engage in the process of learning, our brains undergo a remarkable transformation, allowing for the creation of fresh neural pathways. Consequently, each instance of acquiring knowledge possesses the capacity to reshape the functioning of our brains, enabling us to deviate from our usual cognitive patterns.

The remarkable flexibility of the brain may not be fully harnessed in the acquisition of new information, yet it is certainly evident when mastering a new language or musical instrument. It is clear that not all forms of knowledge are of the same importance. Engaging in such learning experiences could potentially enable individuals to intentionally restructure the brain.

The degree to which we harness the seemingly extraordinary capabilities of the brain is also contingent upon our dedication to fostering neuroplasticity and the manner in which we lead our lives.

1.2 Children's Neuroplasticity

During childhood, neuroplasticity is at its peak as an inherent aspect of human development. It can be perceived as a vital mechanism for children in terms of their ability to adapt to risk and exhibit resilience [6].

The cognitive development of children is an ongoing and dynamic process, marked by constant growth and alterations in the brain. Every new experience plays a role in altering neural connections or reshaping the brain's functionality, and at times, it may have an impact on both aspects simultaneously.

During the early stages of development, a newborn's brain undergoes a remarkable process wherein each neuron establishes around 7500 connections with other neurons. This astonishing number of connections surpasses the quantity observed in the average adult brain by the time the child reaches the age of two [7].

However, as the child grows older and begins to forge their own distinctive patterns and connections, these connections gradually diminish. This phenomenon highlights the presence of neuroplasticity in children, which can be categorized into four distinct groups encompassing a wide range of neural activity: impaired, excessive, adaptable, and plasticity [7].

1.3 Adult Neuroplasticity

Adults retain this ability, albeit with reduced frequency and intensity compared to children. Nonetheless, the adult brain possesses the potential for remarkable transformation despite this. Memory and overall cognitive function can be improved through this process, which can also revive obsolete connections and dormant abilities.

Despite the fact that older individuals often possess less potential compared to children and young adults, adults have the ability to foster positive transformations and development in their brains similar to younger generations through perseverance and adopting a healthy lifestyle. Continue reading to explore several remarkable ways in which neuroplasticity can impact the adult brain!

1.4 Some New Studies on Neuroplasticity

Here are a small number of the more exciting and happenings in the field:

1. Neurons that are in their "newborn" stage typically reach the same level of development

as older neurons by the time they are 8 weeks old (Aditi Deshpande, Matteo Bergami, Alexander Ghanem; "Retrograde monosynaptic tracing reveals the temporal evolution of inputs onto new neurons in the adult dentate gyrus and olfactory bulb," 2013).

2. Engaging in regular physical activity and maintaining physical fitness levels can positively impact the size of the hippocampus. This, in turn, can help to mitigate or reduce the typical age-related decline and deterioration of neurons, according to Niemann et al. (C Niemann, B Godde, C Voelcker-Rehage; "Not only cardiovascular, but also coordinative exercise increases hippocampal volume in older adults," Frontiers in aging neuroscience, 2014).
3. According to Vasconcelos et al. (2014), adequate rest can improve the adaptability of synaptic responses, while persistent insomnia results in the deterioration of neural connections (neuronal death and damage) in the hippocampus (Araujo Lima, Alisson Menezes, Veralice Meireles Sales de Bruin, Emiliano Ricardo Vasconcelos Rios, and Pedro Felipe Carvalhedo de Bruin; "Differential effects of paradoxical sleep deprivation on memory and oxidative stress." Naunyn-Schmiedebergs Archives of Pharmacology 387, no. 5, 2014).
4. The midbrain dorsal raphe nucleus (DR), a key site for most ascending serotoninergic projections, demonstrated notable neuroplasticity following the experience of motherhood. The plasticity observed in the DR could potentially contribute to the neurochemical alterations necessary for effective maternal caregiving, as serotonin plays a vital role in postpartum behaviors and maternal emotional well-being (M. Allie Holschbach, Joseph S. Lonstein; Neuroscience Program, 108 Giltner Hall, Michigan State University, East Lansing, MI 48824, USA, 2016. https://doi.org/10.1016/j.psyneuen.2016).
5. Music can make our brains active and engaged. It stimulates parts of our brain like the amygdala and hippocampus, which are responsible for emotions and memory. It also affects other areas such as the orbitofrontal cortex, parahippocampal gyrus, and temporal poles, which play a role in generating and controlling emotions. Because of this, music might have an impact on how we develop socially and emotionally.

1.5 The Benefits Derived from Rewiring the Brain

By using brain rewiring programs, the limbic system may be updated, stress and chronic inflammation can be decreased, and people can restore their health and wellness.

The concept of neuroplasticity, which demonstrates the brain's capacity to modify and adjust in reaction to novel experiences, serves as the foundation for initiatives aimed at reprogramming the brain. The tasks you complete as part of a brain rewiring program are designed to help your brain create new neural connections that will enhance the nervous system's ability to self-regulate.

In addition to the benefits previously discussed, brain rewiring can help with goal-setting and achievement, increase adaptability and flexibility, enhance cognitive function, increase healing capacity, and offer clarity and inner peace.

2 How the Brain Develops

From the intense expansion and growth in complexity before birth to the refinement and reduction of connections in childhood and adolescence, the brain responds to life experiences and the environment. This process continues into early adulthood, with the brain retaining the capacity to change and adapt at any age. The birth of new neurons, the death of old ones, the creation and dissipation of neural connections, and other internal processes contribute to the brain's ability to continually evolve.

This passage describes the remarkable development and plasticity of the human brain, from pre-birth to adulthood and beyond. Here's a breakdown of the key points:

1. Pre-birth development: The brain undergoes intense growth and expansion before birth, laying the foundation for future development.
2. Early childhood: After birth, the brain continues to grow and develop rapidly, with a "boom period" of increased complexity.
3. Refinement and reduction: As children and adolescents grow, their brains respond to their environment and experiences, strengthening connections between brain cells and pruning less-used ones.
4. Maturity: During early adulthood, the brain generally undergoes significant development and reaches a state of maturity.
5. Ongoing change and plasticity: However, the process of neural development and remodeling does not cease in adulthood. Throughout life, the brain:

   - Continues to generate new neurons (neurogenesis)
   - Eliminates or reorganizes old neurons and connections (apoptosis)
   - Creates new neural connections (synaptogenesis)
   - Dissolves existing connections (synaptic pruning)

In essence, the passage highlights the remarkable capacity of the brain to change and adapt throughout life, emphasizing that it is never too late to learn, grow, and reorganize.

3 Stages of Early Brain Development

Brain development is a complex and dynamic process that occurs at different rates in different regions of the brain. Some areas mature earlier, while others continue to develop into adulthood. This variability is influenced by a combination of genetic factors, environmental experiences, and individual differences.

For example, the prefrontal cortex, which is responsible for higher-order cognitive functions such as decision-making and impulse control, undergoes significant development during adolescence and continues into early adulthood. On the other hand, regions involved in sensory processing or basic motor functions may mature earlier in childhood. Additionally, there can be individual variations in the timing and sequence of brain development. Factors such as genetics, nutrition, exposure to stress or trauma, educational opportunities, and social interactions can all influence how the brain develops.

Overall, this complexity highlights that brain development is not a linear process but one with multiple factors influencing its trajectory. Understanding these variations helps us appreciate the unique developmental journeys individuals experience as they grow from childhood into adulthood.

Different regions of the brain mature at different rates, and periods of plasticity can extend well into adulthood. Furthermore, individual differences in genetic makeup or environmental factors can impact the pace and pattern of brain development.

In conclusion, the process of brain development from childhood into adulthood is complex and influenced by various factors. There are some major trends in the construction and remodeling that take place:

- Rapid growth in the womb: During the last trimester of pregnancy, a fetus's brain undergoes rapid growth and development. This includes increases in brain size, the folding of the cortex (the outer layer of the brain), and the establishment of connections between neurons. The formation of major long-range signaling pathways, composed of white matter, also begins during this time and continues to develop until birth. This period is crucial for laying the foundation for future cognitive and neurological development.
- The first 2 years: Brain development during the first 2 years of life is a critical and rapid process. Here are some key points about brain development during this period:

  1. Formation of neural connections: The brain starts developing even before birth,

but it goes through significant growth and wiring during the first 2 years. Neural connections between different parts of the brain are established, allowing for communication and information processing.

2. Rapid growth: The brain experiences substantial growth in size during this time. By age two, a child's brain reaches about 80% of its adult size.
3. Synaptic pruning: As the brain develops, it goes through a process called synaptic pruning, where unused or unnecessary neural connections are eliminated while important ones are strengthened. This process helps streamline neural pathways and optimize efficiency.
4. Sensory development: During infancy, sensory systems such as vision, hearing, touch, taste, and smell develop rapidly. The stimulation received through these senses plays a crucial role in shaping neural pathways and building cognitive abilities.
5. Language development: Language acquisition is one of the most significant milestones during early brain development. Babies begin to recognize sounds and patterns from their environment shortly after birth and gradually learn to understand language and communicate themselves.
6. Motor skills development: During the first 2 years, babies go through significant motor skill developments, such as lifting their head up when lying on their stomachs (around 3–4 months), rolling over (around 6 months), sitting without support (around 7–8 months), crawling (around 9–10 months), standing with support (around 11–12 months), walking with assistance (around 12–14 months), and eventually walking independently (between 12 and 24 months).
7. Cognitive abilities: Simple cognitive processes like attention span, memory formation, and problem-solving skills start emerging during this period as babies learn to explore their surroundings and interact with objects and people around them.

It is important to note that every child's developmental timeline may vary slightly within these general guidelines due to individual differences in genetics or environmental factors they experience early in life.

- Developmental peaks in childhood: During childhood, the brain undergoes significant development and growth. Neurodevelopmental peaks in childhood refer to periods when certain skills and abilities are rapidly developing. These peaks can vary depending on the specific area of brain development. For example, language skills typically peak around 2–3 years old, while social and emotional development may peak at different stages. It's important to note that every child develops at their own pace, so these peaks are general guidelines rather than strict timelines. Additionally, early childhood experiences play a crucial role in shaping brain development and can have long-lasting effects on cognitive, emotional, and social functioning. Around puberty (approximately age 10 for girls and age 14 for boys), there is a rapid increase in gray matter volume, followed by a pruning process where unnecessary connections are eliminated to increase efficiency. This pruning process continues into early adulthood. Different regions of the brain mature at different rates. Areas involved in movement and sensation, such as the motor cortex and sensory cortex, reach maturity earlier than regions involved in higher-order cognitive functions like executive control. The prefrontal cortex, which plays a crucial role in decision-making, impulse control, and complex reasoning skills associated with executive functions, continues to develop throughout adolescence and into early adulthood. This prolonged development of the prefrontal cortex contributes to changes in judgment and decision-making abilities seen during adolescence. Areas beneath the cortex (such as those within the limbic system) exhibit relatively little change during this period. The limbic system plays a role in emotions, memory formation, and motivation. These developmental processes highlight how different regions of the brain mature at different rates during childhood and adolescence. It's important to consider these developmental trajectories when understanding cognitive abilities across various age groups.

- A transitional brain during adolescence: The reduction in synapses and cortical gray matter thinning are part of the maturation process, allowing the brain to become more efficient and specialized in its functioning. These changes contribute to improvements in cognitive abilities and higher-order thinking skills as teenagers transition into adulthood. Some parts of the cortex (such as the amygdala and hippocampus) increase in volume, while others (such as the striatum) are reduced. The prefrontal cortex, which is responsible for higher-order cognitive functions such as decision-making, impulse control, and social behavior, is still developing during adolescence. This can contribute to characteristics often associated with teenagers, such as increased risk-taking behavior and emotional volatility. The limbic system, which is involved in emotions and motivations, also undergoes changes during adolescence. This can lead to heightened emotional responses and sensitivity to social cues. Additionally, the reward system in the brain develops during adolescence. This can influence a teenager's response to rewards and how they seek out novel experiences. Overall, the adolescent brain is characterized by ongoing development in areas related to decision-making processes while also experiencing increased emotional reactivity. These developmental changes play a role in shaping teenage behavior and cognitive abilities.
- Brain maturity in early adulthood: During early adulthood, the prefrontal cortex, which is responsible for executive functions such as planning, decision-making, and impulse control, continues to undergo maturation. This area becomes more efficient in processing information and integrating complex thoughts. Furthermore, connections between different brain regions become stronger and more specialized during this time. This enhanced connectivity allows for better coordination between different areas of the brain involved in various cognitive processes. Additionally, myelination—the process by which nerve fibers are coated with a fatty substance called myelin—continues throughout early adulthood. Myelin acts as an insulator that helps increase signal speed and efficiency within neural circuits. As a result of increased myelination during this stage of life, communication between neurons becomes faster and more effective. In terms of emotional regulation, there is evidence to suggest that the amygdala—a structure involved in processing emotions—becomes better integrated with other regions responsible for cognitive control during early adulthood. This integration allows individuals to regulate their emotions more effectively and make decisions based on long-term goals rather than immediate gratification.

  Overall, brain maturity in early adulthood signifies the completion or near-completion of major developmental processes within the brain. It involves improvements in executive functions like planning and decision-making, as well as enhanced emotional regulation capabilities. However, it should be noted that brain development is an ongoing process that continues throughout adulthood, with smaller changes occurring over time.

4 Introduction to Brain Creation

The complex organ of the brain is formed throughout the course of several developmental phases that start prior to birth and last into adulthood. The following information explains how a brain develops:

- The neural tube, a specialized area in embryonic development, establishes the basis for the formation of the brain. Originating from the neural tube are the brain and spinal cord,

which begin to develop around 3 weeks after fertilization [8].
- During the fetal period, the brain undergoes rapid expansion, reaching approximately 25% of its adult size by the conclusion of the first trimester and 60% by the end of the second trimester. Following birth, the brain continues to develop and grow, gradually reaching around 80% of its adult size by the age of three [8].
- The billions of "neurons" that make up the brain transmit chemical signals to one another. An axon, a cell body, and dendrites make up the three primary components of a neuron. The nucleus and other organelles are found in the cell body. The axon is a protrusion that extends far from the cell body and conveys messages. Short branches called dendrites receive information from other neurons.

The brain consists of various regions, each with its own specific role. The cerebrum, the largest part of the brain, is in charge of cognitive functions like language, memory, and reasoning. Acting as a bridge between the two hemispheres of the cerebrum is the corpus callosum, an intricate network of fibers. In each hemisphere, there are four lobes: frontal, parietal, temporal, and occipital [8].

Along with controlling respiration, heart rate, and balance, the brain also has additional regions that play important roles in these processes. The "brainstem," which connects the brain and spinal cord, the "cerebellum," which coordinates movement and motor learning, and the "limbic system," which manages emotion, motivation, and memory, are a few of these.

The brain is always changing and adjusting to new information, experiences, and challenges. This ability is called "neuroplasticity." Neuroplasticity allows the brain to form new connections, strengthen existing ones, or eliminate unnecessary ones. It is most active during early infancy but continues throughout life. However, neuroplasticity is most pronounced during early childhood [9].

5 Neural Circuits

Neurons are arranged into circuits that process particular types of information, so they never work independently. A neural circuit consists of interconnected neurons that are linked by synapses and work collectively to carry out specific functions in response to stimulation. It is not that the human nervous system contains 100 billion neurons, but rather that these nerve cells can communicate with one another in a highly interactive collection of neural networks, which makes it such an amazing machine.

Neural circuits are formed by synaptic connections among groups of neurons, enabling them to carry out specific functions when activated [10, 11]. The formation of extensive brain networks is a result of the interconnections among multiple neuronal circuits. Artificial neural networks, on the other hand, were developed drawing inspiration from brain circuits, albeit with notable distinctions.

Neural circuits are like special roads in the brain that help us think and move. They are made up of both functional and anatomical parts. Every day, the brain keeps changing and making new neurons connect with each other [12].

The way people grow and change can be different for everyone. Sometimes, these changes can make us stronger and more adaptable, but they can also make us more vulnerable. This happens at all levels, from our individual brain circuits to who we become as we grow up. Our brain circuits can change and adapt, but there are also times when they are more sensitive to certain experiences and things in our environment. When we think about how our thoughts, feelings, and actions are connected to our brain circuits, it helps to have a typical path to compare to. This way, we can see how things like genetics and tough experiences can affect how our brains develop and possibly lead to mental health issues [12].

There are many different levels of complexity that can be studied in neural circuits, ranging from individual neurons to entire regions of the brain. For instance, the cerebellar circuit is responsible for motor learning and coordination, the central amygdala circuit records reward prediction errors, and the lateral septum circuit plays a role in social trauma and reward. These are just a few examples of the various neural networks that can be explored [13].

Learning induces alterations in neural circuits through various methods that bring about modifications in the structure and strength of synaptic connections between neurons. Hebbian learning, an extensively researched mechanism, proposes that neurons that activate simultaneously forge stronger connections (fire together, connect together). Consequently, the frequent co-activation of presynaptic and postsynaptic neurons can amplify the efficiency of signal transmission [10].

Spike-timing-dependent plasticity (STDP), which is a variant of Hebbian learning, depends on the precise timing of spikes between neurons and serves as an additional mechanism. The synaptic strength can either be augmented or diminished by STDP, depending on whether the presynaptic spike occurs before or after the postsynaptic spike [11].

Neural circuits can learn and make sense of uncertain information by using random connections between neurons. These connections act like hidden variables and help the brain understand and draw conclusions even in difficult situations [11].

There are certain limits to how much a neural circuit can learn. These limits are determined by things like the size of the network, the noise in the connections, how hard the task is, and how well the neurons are connected. If the network is small, it might not be able to learn as much. But if it gets bigger, it can learn faster and more accurately. Having more neurons and synapses can also make a difference in how well the circuit learns and performs tasks [13].

6 How Neural Circuits Formed?

Axons and dendrites, the projections of neurons that link with other neurons through synapses, develop and are guided to build neural circuits. Chemical and electrical impulses are sent from one neuron to another through synapses, or junctions. The differentiation, migration, and survival of neurons, as well as their synaptic selectivity and plasticity, are all influenced by a mix of genetic and environmental variables throughout neural circuit formation. Specific neurons or groups of neurons may be activated or inhibited by neural circuits in response to sensory input, internal conditions, or behavioral output. The brain and nervous system's neural circuits serve as the foundation for the processing and integration of information.

7 Neuronal Development

Throughout development, neurons undergo a complex process called neuronal development, responsible for transmitting and processing information in the brain and nervous system. This intricate process includes the production and organization of specialized cells. Neuronal development consists of four main stages: neurogenesis, cell migration, differentiation, and outgrowth. Neurogenesis involves the creation of new neurons from stem cells or neuroblasts. Cell migration refers to the movement of neurons to their final locations in the brain. Differentiation is the process by which neurons acquire specific features and functions, such as neurotransmitter type and synaptic connections. Outgrowth involves the extension of axons and dendrites by neurons to create neural circuits. Genetic and environmental factors influence the process of neuronal development, regulating the expression of genes and proteins involved in neuronal specification, guidance, and plasticity. Understanding the complexities of neuronal development is crucial for comprehending the normal functioning of the brain and nervous system, as well as for understanding the impact of diseases and disorders on these intricate systems [11].

8 Neuronal Transmission

Neurons utilize electrical and chemical signals to communicate among themselves and with other cells in a process referred to as neural transmission. Neural transmission consists of two main stages: action potentials and synaptic transmission. When neurons are stimulated by a threshold stimulus, they generate electrical impulses known as action potentials, which travel through the axons of the neurons. The occurrence of action potentials is determined by the opening and closing of voltage-gated ion channels in the neuronal membrane. These action potentials exhibit a binary response, either firing at full strength or not firing at all [14]. On the other hand, synaptic transmission is the process by which neurons transmit and receive chemical signals across synapses, which are the gaps between neurons or between neurons and other cells. In order for synaptic transmission to take place, the release of neurotransmitters from the presynaptic neuron and their subsequent binding to receptors on the postsynaptic cell is crucial [15].

The stimulation or inhibition of the postsynaptic cell depends on the type of neurotransmitter and receptor involved. Neural communication is essential for the proper functioning of the nervous system and is also involved in various significant processes, including sensory perception, cognitive function, memory formation, motor coordination, and emotional responses [15, 16].

9 Neurotransmiters

Neurons establish communication with one another and their target cells through the utilization of neurotransmitters, which are chemical messengers. These neurotransmitters are synthesized and discharged from nerve terminals located in the synaptic cleft, the narrow gap between cells. Upon their release, they attach to receptors on the target cell, initiating a range of responses such as excitation, inhibition, or modulation. Neurotransmitters are pivotal in the regulation of numerous physiological processes, encompassing heart rate, appetite, emotional state, sleep patterns, and memory retention. The human body harbors a repertoire of over 100 distinct neurotransmitters, with some of the most significant ones including glutamate, GABA, acetylcholine, dopamine, serotonin, and norepinephrine [17].

Neurotransmitters are crucial for the proper functioning of intricate neurological systems. These chemical messengers are released from synaptic vesicles in the synaptic cleft and have the ability to interact with receptors on the target cell. The specific receptor to which a neurotransmitter binds determines the impact it has on the target cell. The production of neurotransmitters involves the conversion of readily available and abundant precursors, such as amino acids, through a limited number of biosynthetic steps. Amino acids serve as a common source for the synthesis of numerous neurotransmitters. It is worth noting that humans possess a vast array of neurotransmitters, with estimates suggesting over 100 different types. However, the exact number of neurotransmitters remains uncertain. To facilitate classification and categorization, neurotransmitters can be grouped into various categories, including amino acids, peptides, and monoamines, based on specific criteria [18].

Neurotransmitters possess the ability to elicit excitatory, inhibitory, or modulatory responses in the target cell. The response is dictated by the neurotransmitter's interactions with distinct receptors located on the postsynaptic membrane. Through the modulation of transmembrane ion flow, neurotransmitters can heighten (excitatory) or diminish (inhibitory) the probability of the cell generating an action potential. Type I synapses feature receptors that induce excitatory outcomes, whereas Type II synapses feature receptors that induce inhibitory outcomes [19].

In spite of the numerous synapses present in the human brain, they all function by transmitting only two distinct types of signals. These signals exhibit varying appearances and are situated in different locations on the neurons they impact [20]. Every synapse contains receptors that have the ability to alter their functionality. When neurotransmitters attach to these receptors, a series of reactions is initiated, which assists the cell in regulating its operation [21]. The effects of neurotransmitter binding differ based on the particular receptor involved. For instance, it can either heighten or lessen the cell's sensitivity to forthcoming signals. As mentioned before, a neurotransmitter can solely activate a receptor with no additional direct effects. Consequently, the behavior of a neurotransmitter system is dictated by the interconnections between the neurons utilizing it and by the chemical composition of the receptors.

Neuroscientist Henry Hallett Dale initially proposed Dale's Law, which stipulated that each type of neuron releases a single, unique neurotransmitter over its lifespan. However, recent research conducted in the field has taken a stand against this theory. Presently, the prevailing belief among neuroscientists is that most neurons release the same neurotransmitter from their axons. This is why specific neurons are denoted as "dopaminergic" or "serotonergic," signifying the release of dopamine or serotonin, respectively. Nonetheless, recent discoveries have unearthed that neurons are not steadfast when it comes to the neurotransmitters they release. Under certain circumstances, certain cells have the ability to switch the type of neurotransmitter they release, occasionally even generating up to five distinct kinds. This phenomenon is commonly referred to as "neurotransmitter switching" [22].

Certain neurotransmitters are classified by experts as single ions, such as zinc, which are released during synaptic transmission. Additionally, some gaseous molecules like nitric oxide, carbon monoxide, and hydrogen sulfide are also considered neurotransmitters by certain experts. These gases are produced within the neural cytoplasm and rapidly diffuse across the cell membrane into the extracellular fluid and neighboring cells. This diffusion process triggers the generation of second messengers. However, the study of soluble gas neurotransmitters poses challenges due to their fast-acting nature and immediate breakdown, resulting in their existence for only a brief duration of a few seconds [23].

Glutamate, the most prevalent neurotransmitter in the human brain, plays a crucial role in excitatory signaling at the vast majority of synapses. It is responsible for promoting brain cell activity and facilitating processes related to memory formation [24, 25]. Conversely, other neurotransmitters such as GABA and glycine contribute to the inhibition and calming down of brain activity. GABA, in particular, is highly effective in reducing excessive brain activity and is involved in approximately 40% of inhibitory processes [26]. Glycine, on the other hand, is primarily found in the spinal cord, serving its inhibitory functions in that region. While glutamate and GABA are the most prominent neurotransmitters, other transmitters also play significant functional roles, albeit in fewer synapses. For instance, the dopamine system is a primary target for addictive substances like cocaine and amphetamines, while addictive opiate medications mimic the effects of opioid peptides to regulate dopamine levels and produce their desired outcomes.

Dopamine, an important neurotransmitter, plays a critical role in a range of brain functions such as learning, motor control, pleasure sensation, emotional regulation, and executive functions. It has also been associated with psychiatric and neurological disorders [27]. Special neurons in the midbrain produce dopamine, which then travels to different areas of the brain, including the striatum, nucleus accumbens, prefrontal cortex, and limbic system. When it comes to receptors, dopamine binds to D1 through D5 receptors.

Conversely, serotonin is a neurotransmitter that impacts multiple cognitive processes and neural activity. Many psychiatric and neurological drugs target serotonin. Serotonin also affects gastrointestinal processes such as bowel movement, bladder control, and cardiovascular function [28]. Neurons in the raphe nuclei of the brainstem release serotonin, which then spreads to various parts of the brain, including the cortex, hippocampus, amygdala, and hypothalamus. Serotonin binds to receptors known as 5-HT1 through 5-HT7 receptors.

Norepinephrine, a monoamine, is synthesized in the central nervous system and sympathetic nerves. Neurons in the locus coeruleus of the brainstem release norepinephrine, which then travels to different parts of the brain, such as the cortex, thalamus, hippocampus, and amygdala. NE binds to receptors called alpha- and beta-adrenergic receptors [11]. The locus coeruleus in the brain plays a crucial role in norepinephrine signaling. Norepinephrine influences various processes, including stress, sleep, attention, focus, and inflammation. It also helps regulate the responses of the autonomic nervous system [29].

Lastly, histamine is another neurotransmitter that maintains bodily equilibrium, promotes wakefulness, influences eating behavior, and regulates motivation [30].

10 Neural Darwinism

Neural Darwinism is a scientific hypothesis that applies the concept of natural selection to the brain's growth and functioning. It was proposed by Gerald Edelman, an American scientist and Nobel laureate. According to this hypothesis, the brain's neuronal groups are formed and modified through selection processes during both embryonic development and throughout life. These processes generate variation, prevent degeneration, and enhance communication between neuronal groups [11].

To address several inconsistencies in our comprehension of the development, structure, and functional operation of the central nervous system, Edelman formulated the theory of neuronal group selection (TNGS) [28]. A significant hurdle faced by Edelman was elucidating the process of perceptual categorization without resorting to a central observer or presuming a predetermined organization of information in the world. The development of consciousness, as per Darwinism, is thought to occur within the thalamocortical system, acknowledged as the brain's hub of activity [13].

11 Use It or Lose It

The principle of "use it or lose it" posits that if we fail to utilize a skill or talent, it will gradually diminish over time. This principle holds true for both our physical and mental capacities [11]. For example, if we don't exercise, our muscles will weaken and shrink [3]. Similarly, if we don't engage our minds, we may experience cognitive decline and memory loss [11]. The "use it or lose it" concept is based on the idea that our brain and body need stimulation and challenges to maintain their structure and function [11]. Research studies have shown that individuals who are more educated, actively participate in society, and have social connections tend to have better cognitive function and a lower risk of Alzheimer's disease. This study provides evidence suggesting that it is crucial to engage in both mental and physical activities in order to preserve our cognitive and physical capabilities [11].

The idea of "use it or lose it" is not a simple or universal rule. Our brain and body go through changes as time passes, influenced by various factors such as genetics, environment, diet, stress, illness, and injury. Moreover, some skills have more flexibility or resilience than others, which means they can still be improved or restored even after long periods of inactivity or damage [11]. Therefore, it is better to view the "use it or lose it" concept as a general principle rather than a strict rule.

12 Neuron Growth Factors

Neuron-stimulating factors, also called molecules, have the ability to impact the survival, differentiation, and growth of neurons. These factors are made up of neurotrophins, which are released by astrocytes and attach themselves to receptors found on neurons [16].

Various substances, such as brain-derived neurotrophic factor (BDNF), nerve growth factor (NGF), and colony-stimulating factor 1 (CSF1), play crucial roles in neurogenesis. Lifestyle choices like exercise, mental activity, sleep, meditation, and sexual engagement can enhance the production of new neurons by stimulating the release of these substances. BDNF, a member of the neurotrophin family of growth factors, is particularly important in maintaining neuron survival, promoting neurogenesis, ensuring synaptic flexibility, and supporting cognitive function. BDNF interacts with specific receptors on neurons, such as the TrkB receptor, and has implications in neurological disorders like Alzheimer's disease, Parkinson's disease, and depression.

BDNF is crucial for the formation of long-term memories. Although the majority of neurons in the human brain are established during fetal development, there exists a phenomenon known as neurogenesis, which permits the generation of new neurons from neural stem cells in specific regions of the adult brain. Neurotrophins, including BDNF, are proteins that are pivotal in initiating and controlling neurogenesis. BDNF stands out as one of the most potent neurotrophins [14, 18]. Studies on mice lacking the ability to produce BDNF have shown abnormal brain and sensory nerve system development, leading to a shortened lifespan. This underscores the importance of BDNF in promoting healthy neural growth [19]. Furthermore, other significant neurotrophins, such as NT-3, NT-4, and NGF, share structural similarities with BDNF.

The endoplasmic reticulum is responsible for creating BDNF, a protein that is then released by dense-core vesicles. These vesicles bind to carboxypeptidase E (CPE), and it has been suggested that if this binding is disrupted, BDNF cannot be sorted into the vesicles. When BDNF is not properly sorted, it can lead to severe phenotypes in mice, including postnatal death. Other effects of this disruption include losses in sensory neurons, which can affect respiration, coordination, balance, hearing, and taste. Additionally, knockout animals show abnormalities in the cerebellum and an increase in sympathetic neurons. On the other hand, exercise-induced neurogenesis and improvements in cognitive performance have been linked to the production of BDNF [20]. It has been found that various forms of physical activity can significantly increase BDNF production by threefold.

13 Neuroglial Cells

In the nervous system, there are special cells called neuroglial cells that support and protect neurons. These cells have different names, such as astrocytes, oligodendrocytes, microglia, ependymal cells, and Schwann cells. They are also known as glial cells or glia. There are a lot of them in our brain, with about 5–10 glial cells for every neuron or a whopping 350 billion cells in total. Neuroglial cells, which are abundant in our central nervous system, represent the most prevalent cell type.

Glial cells continue to replicate, supporting neurons in terms of energy and structure and ensuring favorable working conditions for them. Unlike neurons, glial cells lack chemical connections and exhibit less electrical activity. Within the central nervous system, four distinct types of glial cells exist: astrocytes, oligodendroglia, ependymal cells, and microglia. These cells vary in appearance, responsibilities, and origins.

In the brain, astrocytes make up nearly half of all cells and are the most prevalent form of glia. Various neuroglial cell functions in brain development include:

- Facilitating the growth and regeneration of neurites by creating a nurturing and supportive environment during development and nerve repair [14].
- Producing myelin, a lipid substance that envelops specific axons and enhances the speed of nerve signal transmission [26].
- Ensuring the neuron's optimal ionic balance and regulating its activity by controlling the uptake of neurotransmitters [31].
- Providing a structural framework for certain aspects of brain development, such as guiding the growth of axons and the formation of synapses.
- Eliminating pathogens and cellular waste from damaged areas or during regular cell turnover through the action of scavenger cells.
- Participating in inflammatory responses and immune surveillance within the neurological system [32].

14 Summary

During the process of human central nervous system development, the generation of neurons occurs through a combination of asymmetric and symmetric divisions of neural stem cells as well as their descendant progenitor cells. The apical-basal polarity, cell-cycle duration, and other epithelial properties of these cells play a crucial role in determining whether they will proliferate or differentiate, as well as the type of division they undergo. One area of inquiry focuses on the changes in these characteristics when radial glial cells emerge from neuroepithelial cells. Understanding this shift is important for unraveling its impact on cell destiny and neurogenesis [33].

The field that explores the intricate structures and functions of the nervous system by studying how cells interact and differentiate is referred to as the cellular biology of neural development. This area of research encompasses various topics related to neural development, for example, the traits and actions displayed by neural stem and progenitor cells, the regulation and mechanisms of cell division, the factors and processes influencing cell migration, differentiation, and maturation, the impact of physical and mechanical forces on brain development, and the molecular and cellular basis of neurological conditions resulting from abnormal neurodevelopment.

Investigations into many facets of brain development and growth include:

1. The origins and precise definitions of neuroepithelial cells, radial glial cells, basal progenitors, and neuroblasts [11].
2. The mechanisms and regulations governing cell proliferation and differentiation, encompassing cleavage plane orientation, cell cycle control, cell fate determination, and the distinction between symmetric and asymmetric division [29].
3. The processes and elements influencing cell migration and integration, including neurotrophic factors, synapse formation, axon guidance, neurite outgrowth, and axon guidance.
4. Tissue tension, differential development, cortical folding, and pattern formation are among the physical and mechanical variables that impact the development of the nervous system.
5. Aberrant neural growth and development give rise to human nervous system disorders like microcephaly, lissencephaly, schizophrenia, and autism spectrum disorder, which have underlying molecular and cellular causes [29].

The field of research that focuses on understanding the chemical compounds and genetic factors that determine the formation and functioning of the brain is referred to as the molecular basis of brain development. This area of study involves investigating various aspects of brain development, such as:

- Exploring the origins and distinct characteristics of neural stem and progenitor cells, which include neuroepithelial cells, radial glial cells, basal progenitors, and neuroblasts.
- Examining the mechanisms and regulations of noncoding RNAs, splicing, epigenetics, and gene expression during brain development.
- The control of cell division, proliferation, migration, and survival is governed by signaling mechanisms and transcription factors.
- Axon development, synapse formation, and circuit construction are regulated by the interactions between molecules and guidance cues.
- The molecular evolution and adaptability of human brain development involve changes in gene regulation, gene duplication, gene loss, and gene innovation [34, 35].

References

1. Schulz R. Encyclopedia of aging: a comprehensive resource in gerontology and geriatrics. New York: Springer; 2006.
2. Gutchess A. Plasticity of the aging brain: new directions in cognitive neuroscience. Science. 2014;346(6209):579–82.

3. Fuchs E, Flügge G. Adult neuroplasticity: more than 40 years of research. Neural Plast. 2014;2014:1.
4. Demarin V, Morović S. Neuroplasticity. Period Biol. 2014;116(2):209–11.
5. Bergland C. How do neuroplasticity and neurogenesis rewire your brain. Psychology Today. 2017.
6. Masten AS. Resilience in children threatened by extreme adversity: frameworks for research, practice, and translational synergy. Dev Psychopathol. 2011;23(2):493–506.
7. Mundkur N. Neuroplasticity in children. Indian J Pediatr. 2005;72:855–7.
8. Brain architecture – center on the developing child at Harvard University. Available from: https://developingchild.harvard.edu/science/key-concepts/brain-architecture/
9. Brain anatomy and how the brain works. Johns Hopkins Medicine. Available from: https://www.hopkinsmedicine.org/health/conditions-and-diseases/anatomy-of-the-brain
10. Neuronal developmentlstages, facts, summary & brain development. Available from: https://brainmadesimple.com/neuronal-development
11. Neuronal development – latest research and news. Nature. Available from: https://www.nature.com/subjects/neuronal-development
12. Mikoshiba K, Lajtha A. Handbook of neurochemistry and molecular neurobiology: neural signaling mechanisms. New York: Springer Science & Business Media; 2009.
13. Neural development. Available from: https://neuraldevelopment.biomedcentral.com/
14. Neurotransmission. Available from: https://en.wikipedia.org/wiki/Neurotransmission
15. Neural transmission – CliffsNotes. Available from: https://www.cliffsnotes.com/study-guides/psychology/psychology/psychology-biological-bases-of-behavior/neural-transmission
16. Neurotransmitters and receptors (article). Khan Academy. Available from: https://www.khanacademy.org/science/biology/human-biology/neuron-nervous-system/a/neurotransmitters-their-receptors
17. Neurotransmitters: types, functions and disorders. Kenhub. Available from: https://www.kenhub.com/en/library/anatomy/neurotransmitters
18. Prasad BV. Examining biological foundations of human behavior. Hershey: IGI Global; 2020.
19. Peters A, Palay SL. The morphology of synapses. J Neurocytol. 1996;25:687–700.
20. Shier D, Butler J, Lewis R. Hole's human anatomy and physiology, chapter 7. New York: McGraw-Hill; 2009.
21. Di Chiara G, Morelli M, Consolo S. Modulatory functions of neurotransmitters in the striatum: ACh/dopamine/NMDA interactions. Trends Neurosci. 1994;17(6):228–33.
22. Webster R. Neurotransmitters, drugs and brain function. Chichester: John Wiley & Sons; 2001.
23. Kodirov SA, Takizawa S, Joseph J, Kandel ER, Shumyatsky GP, Bolshakov VY. Synaptically released zinc gates long-term potentiation in fear conditioning pathways. Proc Natl Acad Sci. 2006;103(41):15218–23.
24. Gross L. "Supporting" players take the lead in protecting the overstimulated brain. PLoS Biol. 2006;4(11):e371.
25. Bowery N, Smart T. GABA and glycine as neurotransmitters: a brief history. Br J Pharmacol. 2006;147(S1):S109–S19.
26. Corbière A, Vaudry H, Chan P, Walet-Balieu M-L, Lecroq T, Lefebvre A, et al. Strategies for the identification of bioactive neuropeptides in vertebrates. Front Neurosci. 2019;13:948.
27. Ko JH, Strafella AP. Dopaminergic neurotransmission in the human brain: new lessons from perturbation and imaging. Neuroscientist. 2012;18(2):149–68.
28. Berger M, Gray JA, Roth BL. The expanded biology of serotonin. Annu Rev Med. 2009;60:355–66.
29. O'Donnell J, Zeppenfeld D, McConnell E, Pena S, Nedergaard M. Norepinephrine: a neuromodulator that boosts the function of multiple cell types to optimize CNS performance. Neurochem Res. 2012;37:2496–512.
30. Passani MB, Panula P, Lin J-S. Histamine in the brain. Front Media SA. 2014;8:64.
31. Lee K. Neuroglial cells: an overview of their physiological roles and abnormalities in mental disorders. Korean J Biol Psychiatry. 2015;22(2):29–33.
32. Simitzi C, Ranella A, Stratakis E. Controlling the morphology and outgrowth of nerve and neuroglial cells: the effect of surface topography. Acta Biomater. 2017;51:21–52.
33. Götz M, Huttner WB. The cell biology of neurogenesis. Nat Rev Mol Cell Biol. 2005;6(10):777–88.
34. Enard W. The molecular basis of human brain evolution. Curr Biol. 2016;26(20):R1109–R17.
35. La Manno G, Siletti K, Furlan A, Gyllborg D, Vinsland E, Mossi Albiach A, et al. Molecular architecture of the developing mouse brain. Nature. 2021;596(7870):92–6.

Impacts of Environmental Exposures on Neurodevelopment

Sina Arabi and Parnian Poursafa

Abstract

The identification of factors contributing to environmental health issues is challenging because of the complex nature of environment-related health problems. Various environmental factors, including air quality, water quality, food safety, and living environments, play a crucial role in health and diseases. Neurodevelopment is one of the most important health topics. In spite of global endeavors to reduce environmental levels of potentially hazardous substances and consequently mitigate their adverse impacts on the human body, many chemical substances are widespread in the environment and frequently exceed recommended levels. In this chapter, we will briefly discuss the potential environmental factors that have adverse effects on neurodevelopment. Our focus will be on long-term consequences that have received less attention in previous studies. We will delve briefly into the detrimental effects of heavy metals, pesticides, endocrine-disrupting chemicals, and air pollution on neurodevelopment and particularly discuss Fetal Alcohol Spectrum disorders, autism spectrum disorder, and attention deficit hyperactivity disorder.

Keywords

Environmental exposure · Neurodevelopmental disorders · Mental health · Heavy metals · Endocrine disruptors · Pesticides

1 Introduction

In point of fact, the identification of the exact factors that contribute to environmental health issues can be arduous. It appears that health problems associated with the environment are very intricate and complex, and it is challenging to pinpoint how environmental factors cause diseases; for instance, different genetic predispositions could lead to discrete outcomes for each factor. The quality of the air we breathe, the water we consume, the food we ingest, and the environments in which we reside and work all play a decisive role in environmental health concerns. Hazardous chemical substances, such as toxic pesticides, heavy metals, and endocrine-disrupting chemicals (EDCs), have been noted to disproportionately affect children and pregnant

S. Arabi
Child Growth and Development Research Center, Research Institute for Primordial Prevention of Non-communicable Disease, Isfahan University of Medical Sciences, Isfahan, Iran

P. Poursafa (✉)
Institute of Cell Biology and Neuroscience, Goethe University, Frankfurt, Germany
e-mail: s0475808@stud.uni-frankfurt.de

© The Author(s), under exclusive license to Springer Nature Singapore Pte Ltd. 2024

R. Kelishadi (ed.), *Nature, Nurture, and Neurodevelopment*,
https://doi.org/10.1007/978-981-97-7433-3_4

women globally. It has been estimated that more than 20% of all premature deaths annually can be attributed to environmental factors [1]. Neurodevelopmental disabilities, such as autism spectrum disorders (ASD), attention deficit hyperactivity disorder (ADHD), and learning and intellectual disabilities, impose a significant burden on the healthcare system across the world. The care for individuals with neurodevelopmental disorders often necessitates multiple services, including family support, education, social services, and healthcare, which can result in substantial costs. For example, it has been projected that the total costs of caring for children with ASD will exceed $450 billion in 2025 in the USA [2]. Unfortunately, the prevalence of neurodevelopmental disorders such as ADHD and ASD seems to be rising. While extensive research has been conducted on the genetic origins of these disorders, less attention has been given to identifying environmental factors. In spite of genetic factors, which are irreversible, environmental factors are modifiable risk factors and can be averted if preventative measures are taken. Therefore, identifying specific environmental factors that increase the risk of neurodevelopmental disorders will be an effective strategy for primary prevention [3]. In this chapter and upcoming subsections, we will briefly discuss the potential environmental factors that have adverse effects on neurodevelopment. Our focus will be on long-term consequences that have received less attention in previous studies. This chapter will briefly delve into Heavy metals, Pesticides, Endocrine-disrupting chemicals, Air pollution, Fetal Alcohol Spectrum Disorders, Autism spectrum disorder, and Attention Deficit Hyperactivity Disorder in separate subsections.

2 Heavy Metals

Despite our best endeavors to reduce the environmental levels of heavy metals and consequently mitigate their adverse impacts on the human body, many metals and metalloids are widespread in the environment and frequently exceed recommended levels. Due to their non-biodegradable nature, their environmental concentrations can gradually accumulate over time, which could lead to cumulative exposures in the human body that can have long-lasting negative impacts [4]. Many of these heavy metals originate from poorly regulated use in agriculture, mining, chemical products, industrial paints, building materials, and fertilizers [5, 6]. Substantial epidemiological evidence exists linking these metals to adverse neurobehavioral effects in children. Among heavy metals, lead, mercury, arsenic, and cadmium have been shown to have significant consequences on human health, especially in children [7]. In the current subsection, we will briefly discuss the association between exposure to heavy metals and their neurodevelopmental impacts.

2.1 Lead

Human exposure to lead dates back to the ancient era when the Romans expanded the use of lead, mainly in mining and as a sweetener and preservative in wines. Subsequently, the addition of lead to gasoline in the 1920s led to widespread environmental contamination and human exposure. However, due to its detrimental consequences on brain development, leaded gasoline was banned in the 1970s [8, 9]. Unfortunately, lead can still be found in various sources throughout a child's environment today, including certain water pipes, toys, and jewelry. Additionally, children living near airports may face lead exposure from aviation gas present in the air and soil [8].

The scientific literature extensively documents the neurotoxicity resulting from exposure to the metal lead. The earliest studies on the adverse consequences of lead exposure primarily focused on IQ levels. Needleman and colleagues, in 1979, demonstrated that children with elevated lead levels in their teeth had been burdened with a significant reduction in IQ test scores, specifically in verbal IQ scores [9, 10]. After this important breakthrough, numerous research studies have shown that even low-level lead exposure is associated with a reduction in IQ scores [11]. As a result, the Centers for Disease Control (CDC) has revised its acceptable blood lead level limit, starting from 60 μg/dl in the 1970s to 3.5 μg/dl currently [12]. This reduction was implemented due to the findings of many studies that blood lead levels below 10 μg/dl are still associated with a decrease in IQ scores, in such a way that an increase in the lifetime average blood lead concentration of 1 μg per deciliter is associated with a decrease of 1.37 IQ points [13].

In addition to the negative impacts of lead on IQ, several studies have reported significant associations between lead exposure and poorer school performance. One example is the finding that children with a blood lead level between 5 and 10 μg/dL had significantly lower reading and mathematics scores, even after adjusting for IQ scores [14]. This finding has been corroborated by several similar surveys [11, 15, 16]. Furthermore, exposure to lead is strongly associated with Conduct Disorder (CD), a common and highly impairing psychiatric disorder that is characterized by antisocial and aggressive behavior [17, 18]. Consequently, the increase in environmental lead levels over the past decades has been correlated with a similar escalation in intellectual disabilities and violent crimes [19]. However, it is of note that the association between childhood blood lead levels and criminal offending is controversial, and this association was contradicted in a newer study [20]. It is also established that higher levels of childhood lead exposure can increase the risk of Attention Deficit Hyperactivity Disorder (ADHD) amongst children. Please refer to the ADHD subsection in this chapter for more information.

2.2 Mercury

Mercury is an element and a heavy metal that exists in three forms: (i) elemental, (ii) inorganic, and (iii) organic. Elemental mercury can be found in thermometers, mining processes, and dental amalgams. Elemental mercury and inorganic mercury, which is a combination of elemental mercury with other elements such as chlorine and sulfur, are emitted into the environment by both natural and industrial processes, such as volcanoes and burning fossil fuels, respectively. Consequently, methylmercury, which is the organic form of mercury, is formed from inorganic mercury by microorganisms that exist in both water and soil. The entry of methylmercury into the food chain can easily expose people to this toxic substance and additionally lead to its accumulation in the body when they consume fish or shellfish contaminated with methylmercury [21].

The early evidence of methylmercury-induced neurotoxicity dates back to the 1950s in Japan. The pollution of Minamata Bay with methylmercury led to the poisoning of local residents, including pregnant women who consumed large amounts of seafood. As a result, newborn infants born to these mothers developed a variety of neurodevelopmental deficits, now called Minamata Disease. The symptoms of Minamata disease include sensory disorders, ataxia, hearing loss, impaired speech, eye movement disorders, and cerebral palsy [22, 23].

When the harmful effects of methylmercury were discovered in the Minamata case, research teams around the world conducted studies to find out the potential detrimental effects of low-level methylmercury exposure on the human body. First and foremost, it has been demonstrated that methylmercury can transport across both the blood-brain barrier (BBB) and the blood-placental barrier (BPB) [24, 25]. Moreover, a cohort study conducted in the Faroe Islands, which investigated the potential effects of prenatal exposure to methylmercury on school-aged children, revealed that higher concentrations of mercury in mothers' blood and hair were associated with poorer performance in language, attention, memory, visuospatial, and motor function tests amongst the children [26]. Furthermore, similar prospective studies conducted in Canada, New Zealand, the Seychelles Islands, and New Bedford city have shown that even low-level methylmercury exposure can be associated with poorer school performance, lower IQ, memory and learning impairments, and generally, prenatal exposure to low-level methylmercury can disrupt normal developmental processes in children [22, 27–30]. It is also established that methylmercury exposure can increase the risk of ADHD. Please refer to the ADHD subsection in this chapter for more information.

2.3 Arsenic

Millions of people are exposed to elevated levels of inorganic arsenic, which can lead to long-term and short-term detrimental consequences, including diabetes, cancer, neuropathy, and neurodevelopmental diseases. The main source of human exposure to arsenic is contaminated groundwater with high levels of arsenic, in such a way that the World Health Organization (WHO) estimates that about 200 million people are drinking water containing arsenic levels exceeding 10 μg/L. Additionally, natural activities such as volcanic eruptions, as well as human activities like mining, agriculture, and burning fossil fuels, can expose people to dangerous levels of inorganic arsenic [31, 32].

The damaging effects of inorganic arsenic on neurodevelopment were initially discovered in Japan in 1955 when infants were fed with contaminated powdered milk, resulting in over 12,000 cases of poisoning and 130 deaths. After this appalling tragedy, a couple of follow-up studies were conducted on this population, showing that exposure to high levels of arsenic is associated with lower levels of IQ, white matter degeneration, an increased risk of epilepsy, and intellectual disability [31, 33].

In 2007, a monumental study conducted in Mexico, including approximately 550 children aged 6–8, showed a significant inverse association between urinary arsenic levels exceeding 50 μg/L and visual-spatial abilities, language, memory, intelligence, and mathematical skills. These findings were concluded following adjustments for hemoglobin concentration, blood lead concentration, and socioeconomic status [34]. Moreover, a systematic review and meta-analysis published in 2013 revealed that a 50% increase in arsenic levels in urine (ranging from 35 to 117 μg/L) was significantly associated with a decrease of 0.39 points in the full-scale IQ of children aged 5–15 years. Furthermore, while a decrease of 0.26 points was measured in verbal IQ, no significant result was observed in performance IQ [35]. The deleterious impact of arsenic on IQ has been corroborated by several surveys [36].

Despite concrete evidence showing impaired cognition as a result of postnatal exposure to arsenic, the detrimental effects of prenatal exposure to arsenic remain controversial. A large population-based study investigating infants born to mothers with high urinary arsenic levels (median of 81 μg/L) in Bangladesh failed to find an association between impaired cognition and prenatal exposure to arsenic [37]. However, in a follow-up study, this research team discovered the adverse effects of arsenic exposure on IQ at 5 years of age [38]. It has been suggested that the underlying mechanism of arsenic toxicity is that arsenic alters the signaling of essential neurotransmitters such as glutamate, dopamine, serotonin, and gamma-aminobutyric acid (GABA). Additionally, arsenic alters DNA methylation and histone modifications, leading to epigenetic changes [19, 36].

2.4 Others

Human exposure to neurotoxic metals has raised concerns in recent decades. While many studies have shown the detrimental effects of lead, mercury, and arsenic on neurodevelopment, the long-lasting neurotoxicity of other heavy metals such as aluminum, manganese, and copper has been less extensively investigated [39]. In the upcoming paragraphs, I will briefly discuss the harmful impacts of these heavy metals on neurodevelopment.

Regarding aluminum, it is worth mentioning that aluminum is one of the most prevalent metals in the environment, in such a way that millions of people are being exposed to this element through activities such as mining or the consumption of polluted food. Additionally, minor exposures can occur through drinking contaminated water or inhaling low levels of aluminum particles in the air. Studies have indicated that high levels of aluminum could increase the risk of neurodegenerative disorders such as Alzheimer's disease, the most prevalent form of dementia, and Parkinson's disease, a movement disorder of the nervous system [40]. These findings were corroborated in 2020, when it was indicated that individuals who had suffered from Alzheimer's disease, autism spectrum disorder (ASD), and multiple sclerosis had significantly higher levels of aluminum content in their brain tissue [41].

In contrast to aluminum, manganese is an essential trace element for the human body which plays an important role as a coenzyme in several physiological processes. Exposure to high levels of manganese can lead to a neurological condition called manganism, which displays symptoms similar to those of Parkinson's disease. Moreover, elevated levels of manganese can be associated with neurodevelopmental disorders in children, such as reduced IQ levels and impaired verbal learning [39, 42].

Similar to manganese, copper plays a decisive role in certain enzymes as a cofactor. The precise role of copper in the development of neurodegenerative diseases is controversial. While some studies have shown that high levels of copper can lead to neurodegenerative diseases like Alzheimer's and Parkinson's, others have indicated that copper deficiency may be associated with a higher risk of these diseases [43]. Nevertheless, it has been shown that children suffering from autism had significantly higher levels of copper in their hair and nail samples compared to healthy controls [44].

3 Pesticides

Pesticides are widely used substances that consist of various types, including disinfectants, fumigants, fungicides, herbicides, rodenticides, and insecticides. There are more than a thousand pesticides that have been utilized, resulting in the annual usage of approximately 5.5 billion pounds of pesticides, globally. Additionally, about 1 billion people are directly involved in the agricultural sector; hence, we can confidently assert that all individuals throughout the planet, whether directly or indirectly, are being exposed to pesticides, specifically insecticides and herbicides [45, 46]. One of the first modern synthetic pesticides that was extensively used is Dichlorodiphenyltrichloroethane (DDT), which is an insecticide used in agriculture commonly from the 1940s to 1970s. Due to the numerous detrimental consequences of DDT on the environment and human health, over 100 countries fully banned its use in 2001 and signed the 2001 Stockholm Convention [47]. For instance, it has been demonstrated that in utero exposure to DDT can be associated with psychomotor and mental developmental issues [48].

Organochlorine pesticides (OCPs) are a group of chlorinated compounds that appertain to persistent organic pollutants (POPs), which are toxic substances having adverse effects on human health and the environment globally. Several studies have demonstrated that OCPs, such as DDT, Lindane, Dicofol, Dieldrin, and Endosulfan, can cross the placenta and accumulate in breast milk [49, 50]. A survey conducted in 2007, evaluating 465 children with Autism Spectrum Disorders (ASD) and approximately 7000 controls, depicted that in utero exposure to organochlorine pesticides, particularly Dicofol and Endosulfan, significantly increased the risk of developing ASD (OR: 6.1, CI: 2.4–15.3) [51]. It has been suggested that the damaging effects of OCPs are related to alterations in the levels of neurotransmitters [50]. This idea has been corroborated by numerous animal studies. For example, research has shown that prenatal exposure to Dieldrin can lead to alteration in the dopaminergic system in mice [52]. Another study conducted in Guadeloupe, by the assessment of approximately 150 infants, revealed that prenatal and postnatal exposure to low levels of Chlordecone, an insecticide, is associated with cognitive and motor impairments amongst infants [53].

Organophosphorus compounds (OPs) are chemical substances mainly utilized as pesticides, predominantly as herbicides and insecticides; however, OPs are one of the constituents of nerve agents, such as Tabun, which was initially developed as a pesticide but later repurposed as a chemical warfare agent. It has been estimated that about 80 million pounds of OPs are used annually in the USA, leading to significant exposure for individuals, mostly through the consumption of contaminated foods. Furthermore, agricultural laborers and even households using OPs could be prone to higher levels of exposure to these chemicals compared to the general population. Acute exposure to high levels of OPs can result in symptoms such as nausea, vomiting, paralysis, and seizures [54, 55]. In the upcoming paragraphs, we will focus on the long-term deleterious effects of OPs on neurodevelopment.

A meta-analysis of 14 studies evaluating long-term exposure to low levels of OPs showed that OPs aggravated cognitive performance. Specifically, after analyzing data from approximately 1600 individuals, this study revealed that even low-dose exposure to OPs had detrimental effects on cognitive performance, including psychomotor speed, visuospatial ability, working memory, and visual memory [56]. A monumental cohort study conducted in New York, initiated in 1997 with the aim of evaluating prenatal exposure to OPs (specifically Chlorpyrifos), has shown that high levels of Chlorpyrifos in umbilical cord blood and maternal blood (chlorpyrifos levels exceeding 6.17 pg/g) were associated with impaired psychomotor and mental development during the first 3 years of life. Furthermore, this study demonstrated that highly exposed children were prone to develop ADHD [57]. Another study conducted in Ecuador, which examined 84 children (with 35 exposed to pesticides during pregnancy), revealed that prenatal exposure to OPs can be inversely associated with motor coordination, visual memory, and visuospatial performance [58]. However, it is of note that despite evidence illustrating a strong association between prenatal exposure to OPs and cognitive impairment [59], the statistically significant association between urinary levels of OPs and impaired cognitive development is controversial, with some studies finding no correlation [60].

Epidemiological evidence on the association between pesticide exposure and the development of neurodegenerative diseases, such as Alzheimer's and Parkinson's disease, is controversial. Conflicting results have been found in studies; while some studies have shown a statistically significant association, others have not. Further research is needed to illustrate a concrete association. It is worth mentioning that age is a decisive factor in neurodegenerative diseases, and simultaneously, it allows for the lifelong accumulation of chemical substances. Moreover, the majority of studies have been case-control studies relying on self-reported questionnaires, which could lead to bias [61, 62].

4 Endocrine-Disrupting Chemicals

Endocrine-disrupting chemicals (EDCs) are chemical substances, whether natural or artificial, that mimic or interfere with the actions of the body's hormones. People are mostly exposed to EDCs through air, food, and personal care products. In recent years, numerous studies have concretely demonstrated the harmful effects of EDCs on human health, not the least of which is the carcinogenic effect [63]. In the preceding subsections, we have discussed the detrimental effects of some EDCs, such as heavy metals and pesticides, on neurodevelopment. In the upcoming paragraphs, we will briefly discuss the potential noxious impacts of other well-known EDCs on neurodevelopment.

Considering the fact that EDCs can cross the blood-placental barrier (BPB) and the blood-brain barrier (BBB), both prenatal and postnatal exposure to EDCs have been significantly associated with adverse effects on neurodevelopment [64]. Polychlorinated biphenyls (PCBs) are synthetic organic chemicals that had been widely used as coolants, plasticizers, and in various electrical equipment before being banned in the USA in the year 1977. There is evidence that notably depicts impaired learning, memory, and verbal skills due to prenatal and postnatal exposure to PCBs [65]. Numerous large-scale studies conducted around the globe in locations such as Taiwan, the USA, the Netherlands, and Germany have demonstrated the association between prenatal exposure to PCBs and impaired cognitive function amongst infants and children [66]. The precise underlying mechanisms by which PCBs impact neurodevelopment remain vague; however, findings from in vitro and in vivo studies have suggested that PCBs may disrupt the dopaminergic and serotonergic systems. Like many other EDCs, PCBs have been shown to alter thyroid hormone levels, which are decisive for normal brain development [67].

With regard to the adverse effects of phthalates, which are chemicals commonly used as plasticizers and are found in various products such as vinyl flooring, toys, food packaging, and cosmetics, exposure to phthalates mainly occurs through ingestion, inhalation, and hand-to-mouth contact, posing a greater risk to children due to their behavior. Having entered the body, phthalates are metabolized and excreted through urine. A prospective cohort study conducted in Taiwan evaluating the potential detrimental effects of prenatal and postnatal phthalates on neurodevelopment revealed that higher urinary levels of phthalates in children are inversely associated with cognitive development and IQ levels. However, statistically significant associations between maternal phthalate exposure during pregnancy and the children's IQ scores were not found [68]. Furthermore, it has been shown that prenatal and early childhood exposure to phthalates can lead to behavioral and emotional disorders in childhood [69, 70].

Bisphenol A (BPA) is mainly used in the production of plastic materials and is usually found in various products, such as food containers and plastic bottles. Exposure to BPA mostly occurs through the consumption of food or water stored in BPA-containing containers. Additionally, workers in manufacturing industries that use BPA are also at potential risk of high-level exposure. While the health effects of low-level BPA exposure in humans are not fully understood, some human and animal studies suggest that the impacts of BPA on health are sex-dependent [65]. Whereas some studies have depicted an association between prenatal and early childhood exposure to BPA and impaired cognitive function and IQ levels, this association was not found in other surveys [71, 72]. However, a noteworthy study evaluating more than 450 children showed that higher urinary levels of BPA were significantly associated with the prevalence of ADHD, especially amongst boys [73].

The majority of studies evaluating the deleterious impacts of EDCs on neurodevelopment have focused on the effects of pesticides, heavy metals, PCBs, phthalates, and BPA. However, there are also studies that have investigated other EDCs, such as Polycyclic Aromatic Hydrocarbons (PAHs) and polyfluoroalkyl substances (PFAS). For instance, regarding PAHs, which are commonly found in petroleum-based products, research has shown that prenatal exposure to high levels of PAHs can result in a significant reduction in children's IQ levels at the age of 5 [74]. Similar to PAHs, PFAS are widely used in industry in such a way that millions of people are being exposed to it directly or indirectly, and surprisingly, PFAS has been detected in 97% of serum samples from Americans [75]. A notable cohort study conducted in the USA has shown that even prenatal exposure to low levels of PFAS can increase the risk of ASD in children [76].

It is estimated that people are being exposed to approximately 1000 EDCs [75]; although extensive research has been conducted on the potential adverse effects of these EDCs on neurodevelopment, further investigation is indisputably required, and measures should be taken to rectify the possible adverse effects.

5 Air Pollution

Air pollution refers to the presence of harmful particles or excessive quantities thereof in the air that can lead to detrimental effects on human health, the environment, and the whole ecosystems in general. Common sources of air pollution include emissions from vehicles, industrial/agricultural activities, and natural resources such as volcanic eruptions. Numerous adverse effects of hazardous levels of air pollution on human health have been discovered in recent decades, such as respiratory diseases (especially asthma), cardiovascular problems, and even premature death. It has been estimated that over 5 million deaths are attributable to ambient air pollution annually [77].

The main substances that contribute to air pollution are PAHs, heavy metals, carbon monoxide, and sulfur dioxide. For more information about PAHs and heavy metals, please refer to the subsections on endocrine-disrupting chemicals and heavy metals, respectively. It has been demonstrated that long-term exposure to particles smaller than 2.5 micrometers (PM2.5) and ozone (O3) can be associated with the development of Alzheimer's disease commencing at young ages [78]. A noteworthy study conducted in Spain, which evaluated over 250 children aged 8 to 12 years, showed that exposure to air pollution, even without any structural alterations, could have adverse effects on normal brain development [79]. Furthermore, it has been indicated that prenatal and early life exposure to traffic-related air pollution, including nitrogen dioxide (NO2), PM2.5, and PM10, is significantly associated with the development of ASD [80]. Comparably, it has been illustrated that in utero exposure to air pollution (PM2.5 and PM10), particularly in the last 3 months of pregnancy, is associated with impaired motor skills, cognition, and language development [81]. These findings have been corroborated later by subsequent studies [82].

While further studies are needed to find out the long-term effects of air pollution on neurodevelopment, we concretely assert that measures should be taken to avert the probable noxious consequences of exposure to high levels of air pollution, particularly in developing countries where a larger population is affected by polluted air. Reducing air pollution necessitates a comprehensive strategy to be devised by leading experts; however, one effective approach would be the expansion of cleaner energy resources to reduce levels of PAHs, NO2, PM2.5, and PM10 in the air.

6 Fetal Alcohol Spectrum Disorders

Although human exposure to ethanol dates back to ancient times, the adverse effects of alcohol consumption during pregnancy were not documented until 1968, when a French research team demonstrated that prenatal alcohol exposure could result in malformations, psychomotor anomalies, and intellectual disabilities in children [83]. This monumental study brought to light the potential noxious consequences of alcohol drinking during pregnancy, which is now recognized as Fetal Alcohol Spectrum Disorders (FASDs).

FASDs encompass a range of physical and mental symptoms that infants and children may experience as a result of maternal alcohol consumption during pregnancy. The prevalence of FASDs varies across countries and communities, ranging from 0.01% to 5%. FASDs is a general term used to describe a spectrum of physical, mental, behavioral, and cognitive impairments that can occur in individuals exposed to alcohol prenatally. Different FASD diagnoses, based on the degrees of severity, include Fetal Alcohol Syndrome (FAS), Alcohol-Related Neurodevelopmental Disorder (ARND), Alcohol-Related Birth Defects (ARBD), and Neurobehavioral Disorder Associated with Prenatal Alcohol Exposure (ND-PAE) [84, 85].

Pathological and neuroimaging techniques have depicted that in utero alcohol exposure alters brain structure, such as reductions in brain volume, particularly in the frontal lobe, thalamus, hippocampus, cerebellum, striatum, and caudate nucleus [86–90]. There are several factors that determine the severity of symptoms, not the least of which is exposure time. The critical period during which alcohol interferes with normal development is the first trimester, which is crucial for organogenesis. But by the same token, exposure during the second and third trimesters can result in neurological defects of lesser severity [91]. Consequently, the Centers for Disease Control and Prevention (CDC) asserts that "there is no known safe amount of alcohol consumption during pregnancy. Additionally, there is no safe time to drink during pregnancy, and a woman should avoid alcohol if she is pregnant or may become pregnant" [92].

To establish a FASD diagnosis, three main criteria should be present: (i) growth retardation, (ii) central nervous system (CNS) abnormalities, and (iii) a distinct facial appearance. Regarding facial features, short palpebral fissures, a thin vermillion border, and a smooth philtrum are the major characteristics of a child with FASD [93].

Involvement of the central nervous system (CNS) in children with FASD can lead to a variety of neurodevelopmental and behavioral disabilities that may not become completely apparent until the child reaches school age. The clinical manifestations of these disabilities can vary among patients. Numerous researches have shown that children with FASD tend to score lower on verbal skills compared to unaffected individuals. Additionally, prenatally alcohol-exposed children display reduced problem-solving and planning abilities when compared to control peers. Furthermore, children with FASD commonly experience deficits in language skills, including articulation and grammatical ability. It is notable that FASD cases typically have lower overall IQ scores than the general population. Other atrocious symptoms include impaired memory and motor function, social skills, and sleep patterns [94–96].

7 Autism Spectrum Disorder

Autism spectrum disorder (ASD) is a neurodevelopmental condition that can result in impairments in social communication and interaction. Globally, approximately 1% of children are diagnosed with ASD. Research has indicated that certain environmental pollutants such as lead, methylmercury, PCBs, pesticides, and EDCs may increase the risk of ASD [97, 98]. Studies have shown that methylmercury and lead are significant risk factors for ASD; while exposure to inorganic mercury was significantly associated with an increased risk of ASD and cognitive impairment, exposure to mercury through vaccination, specifically Thimerosal, which contains ethyl mercury, has not been statistically concluded to increase the risk of autism. The association between aluminum exposure and ASD risk is still controversial [99–102].

Exposure to high levels of air pollution has been associated with an increased risk of ASD, especially when exposure occurs during the final trimester of pregnancy. Prenatal and early life exposure to traffic-related air pollution, including nitrogen dioxide (NO2), PM2.5, and PM10, has been significantly associated with the development of ASD [80].

While some studies suggest that prenatal exposure to pesticides increases the risk of ASD, others have not found a significant association. Notable research has shown that even low levels of PFAS exposure during pregnancy can escalate the risk of ASD amongst children. PCBs have also been identified as potential risk factors for ASD [76, 103].

Finally, consumption of foods and drugs can also be a determinant in developing ASD; for instance, inadequate omega-3 intake during pregnancy could be associated with an increased risk of ASD [104]. Additionally, a monumental study conducted in Denmark from 1996 to 2006 revealed a significant association between maternal use of valproate during pregnancy and the risk of ASD [105].

8 Attention Deficit Hyperactivity Disorder

Attention Deficit Hyperactivity Disorder (ADHD) is a neurodevelopmental disorder characterized by symptoms of hyperactivity, impulsivity, and inattention. The prevalence of ADHD among children is estimated to be between 5% and 8% [106]. A remarkable study conducted in Sweden, which evaluated approximately 1 million children, demonstrated a strong association between maternal smoking during pregnancy and ADHD in the offspring [107]. Additionally, alcohol consumption has also been associated with an increased risk of ADHD in children [108]. Conversely, no concrete association has been found between prenatal caffeine consumption and the development of ADHD. Regarding the use of drugs during pregnancy and the risk of ADHD, a large-scale study involving about 38,000 families revealed that children of mothers treated with bupropion during pregnancy have an elevated risk of being diagnosed with ADHD [109]. While it has been suggested that exposure to lead may be associated with an increased risk of ADHD and lower IQ levels, the evidence regarding mercury and manganese remains controversial [110].

9 Conclusion

Evidently, there are a plethora of environmental factors that have detrimental effects on neurodevelopment. The majority of studies have demonstrated that prenatal and early life exposure to many chemical substances is associated with tremendously negative outcomes. Heavy metals, pesticides, air pollution, endocrine disruptors, and alcohol all play a crucial role in impaired neural development.

References

1. Zarocostas J. Millions of deaths from environmental causes are preventable, says WHO. BMJ (Clinical research ed). 2006;332(7555):1412.
2. Leigh JP, Du J. Brief report: forecasting the economic burden of autism in 2015 and 2025 in the United States. J Autism Dev Disord. 2015;45(12):4135–9.
3. Kularatna S, Jadambaa A, Senanayake S, Brain D, Hawker N, Kasparian NA, et al. The cost of neurodevelopmental disability: scoping review of economic evaluation methods. Clinicoecon Outcomes Res. 2022;14:665–82.
4. Heng YY, Asad I, Coleman B, Menard L, Benki-Nugent S, Hussein Were F, et al. Heavy metals and neurodevelopment of children in low and middle-income countries: a systematic review. PLoS One. 2022;17(3):e0265536.
5. Falck AJ, Mooney S, Kapoor SS, White KM, Bearer C, El Metwally D. Developmental exposure to environmental toxicants. Pediatr Clin N Am. 2015;62(5):1173–97.
6. Hsueh YM, Lee CY, Chien SN, Chen WJ, Shiue HS, Huang SR, et al. Association of blood heavy metals with developmental delays and health status in children. Sci Rep. 2017;7:43608.
7. Bauer JA, Fruh V, Howe CG, White RF, Henn BC. Associations of metals and neurodevelopment: a review of recent evidence on susceptibility factors. Curr Epidemiol Rep. 2020;7(4):237–62.
8. Cory-Slechta DA. Chapter 7 – developmental exposure to lead: overview and integration of neurobehavioral consequences and mediation. In: Aschner M, Costa LG, editors. Environmental factors in neurodevelopmental and neurodegenerative disorders. Boston: Academic Press; 2015. p. 139–65.
9. Needleman H. Lead poisoning. Annu Rev Med. 2004;55:209–22.
10. Needleman HL, Gunnoe C, Leviton A, Reed R, Peresie H, Maher C, et al. Deficits in psychologic and classroom performance of children with elevated dentine lead levels. N Engl J Med. 1979;300(13):689–95.
11. Bellinger DC. Very low lead exposures and children's neurodevelopment. Curr Opin Pediatr. 2008;20(2):172–7.
12. Paulson JA, Brown MJ. The CDC blood lead reference value for children: time for a change. Environ Health. 2019;18(1):16.
13. Canfield RL, Henderson CR Jr, Cory-Slechta DA, Cox C, Jusko TA, Lanphear BP. Intellectual impairment in children with blood lead concentrations below 10 microg per deciliter. N Engl J Med. 2003;348(16):1517–26.
14. Surkan PJ, Zhang A, Trachtenberg F, Daniel DB, McKinlay S, Bellinger DC. Neuropsychological function in children with blood lead levels <10 microg/dL. Neurotoxicology. 2007;28(6):1170–7.
15. Skerfving S, Löfmark L, Lundh T, Mikoczy Z, Strömberg U. Late effects of low blood lead concentrations in children on school performance and cognitive functions. Neurotoxicology. 2015;49:114–20.
16. Wang CL, Chuang HY, Ho CK, Yang CY, Tsai JL, Wu TS, et al. Relationship between blood lead concentrations and learning achievement among primary school children in Taiwan. Environ Res. 2002;89(1):12–8.
17. Braun JM, Froehlich TE, Daniels JL, Dietrich KN, Hornung R, Auinger P, et al. Association of environmental toxicants and conduct disorder in U.S. children: NHANES 2001–2004. Environ Health Perspect. 2008;116(7):956–62.
18. Rauh VA, Margolis AE. Research review: environmental exposures, neurodevelopment, and child mental health – new paradigms for the study of brain and behavioral effects. J Child Psychol Psychiatry. 2016;57(7):775–93.
19. Heyer DB, Meredith RM. Environmental toxicology: sensitive periods of development and neurodevelopmental disorders. Neurotoxicology. 2017;58:23–41.
20. Beckley AL, Caspi A, Broadbent J, Harrington H, Houts RM, Poulton R, et al. Association of childhood blood lead levels with criminal offending. JAMA Pediatr. 2018;172(2):166–73.
21. Rice KM, Walker EM Jr, Wu M, Gillette C, Blough ER. Environmental mercury and its toxic effects. J Prev Med Public Health. 2014;47(2):74–83.
22. Burbacher TM, Ponce R, Grant KS. Chapter 6 – the role of methylmercury exposure in neurodevelopmental and neurodegenerative disorders. In: Aschner M, Costa LG, editors. Environmental factors in neurodevelopmental and neurodegenerative disorders. Boston: Academic Press; 2015. p. 107–37.
23. Yorifuji T. Lessons from an early-stage epidemiological study of Minamata disease. J Epidemiol. 2020;30(1):12–4.

24. Kajiwara Y, Yasutake A, Adachi T, Hirayama K. Methylmercury transport across the placenta via neutral amino acid carrier. Arch Toxicol. 1996;70(5):310–4.
25. Kerper LE, Ballatori N, Clarkson TW. Methylmercury transport across the blood-brain barrier by an amino acid carrier. Am J Phys. 1992;262(5 Pt 2):R751–5.
26. Grandjean P, Weihe P, White RF, Debes F, Araki S, Yokoyama K, et al. Cognitive deficit in 7-year-old children with prenatal exposure to methylmercury. Neurotoxicol Teratol. 1997;19(6):417–28.
27. Boucher O, Jacobson SW, Plusquellec P, Dewailly E, Ayotte P, Forget-Dubois N, et al. Prenatal methylmercury, postnatal lead exposure, and evidence of attention deficit/hyperactivity disorder among Inuit children in Arctic Québec. Environ Health Perspect. 2012;120(10):1456–61.
28. Jacobson JL, Muckle G, Ayotte P, Dewailly É, Jacobson SW. Relation of prenatal methylmercury exposure from environmental sources to childhood IQ. Environ Health Perspect. 2015;123(8):827–33.
29. Marsh DO, Clarkson TW, Myers GJ, Davidson PW, Cox C, Cernichiari E, et al. The Seychelles study of fetal methylmercury exposure and child development: introduction. Neurotoxicology. 1995;16(4):583–96.
30. Orenstein ST, Thurston SW, Bellinger DC, Schwartz JD, Amarasiriwardena CJ, Altshul LM, et al. Prenatal organochlorine and methylmercury exposure and memory and learning in school-age children in communities near the New Bedford Harbor Superfund site. Massachusetts Environ Health Perspect. 2014;122(11):1253–9.
31. Dakeishi M, Murata K, Grandjean P. Long-term consequences of arsenic poisoning during infancy due to contaminated milk powder. Environ Health. 2006;5:31.
32. Naujokas MF, Anderson B, Ahsan H, Aposhian HV, Graziano JH, Thompson C, et al. The broad scope of health effects from chronic arsenic exposure: update on a worldwide public health problem. Environ Health Perspect. 2013;121(3):295–302.
33. Yamashita N, Doi M, Nishio M, Hojo H, Tanaka M. Recent observations of Kyoto children poisoned by arsenic tainted "Morinaga Dry Milk" (author's transl). Nihon Eiseigaku Zasshi. 1972;27(4):364–99.
34. Rosado JL, Ronquillo D, Kordas K, Rojas O, Alatorre J, Lopez P, et al. Arsenic exposure and cognitive performance in Mexican schoolchildren. Environ Health Perspect. 2007;115(9):1371–5.
35. Rodríguez-Barranco M, Lacasaña M, Aguilar-Garduño C, Alguacil J, Gil F, González-Alzaga B, et al. Association of arsenic, cadmium and manganese exposure with neurodevelopment and behavioural disorders in children: a systematic review and meta-analysis. Sci Total Environ. 2013;454–455: 562–77.
36. Masjosthusmann S, Tigges J, Fritsche E, Koch K. Chapter two – arsenic-mediated developmental neurotoxicity: recent advances in understanding the adverse outcomes and underlying mechanisms. In: Aschner M, Costa LG, editors. Advances in neurotoxicology, vol. 5. Amsterdam: Academic Press; 2021. p. 35–80.
37. Tofail F, Vahter M, Hamadani JD, Nermell B, Huda SN, Yunus M, et al. Effect of arsenic exposure during pregnancy on infant development at 7 months in rural Matlab. Bangladesh Environ Health Perspect. 2009;117(2):288–93.
38. Hamadani JD, Tofail F, Nermell B, Gardner R, Shiraji S, Bottai M, et al. Critical windows of exposure for arsenic-associated impairment of cognitive function in pre-school girls and boys: a population-based cohort study. Int J Epidemiol. 2011;40(6): 1593–604.
39. Tshala-Katumbay D, Mwanza J-C, Rohlman DS, Maestre G, Oriá RB. A global perspective on the influence of environmental exposures on the nervous system. Nature. 2015;527(7578):S187–S92.
40. Kumar V, Gill KD. Aluminium neurotoxicity: neurobehavioural and oxidative aspects. Arch Toxicol. 2009;83(11):965–78.
41. Exley C, Clarkson E. Aluminium in human brain tissue from donors without neurodegenerative disease: a comparison with Alzheimer's disease, multiple sclerosis and autism. Sci Rep. 2020;10(1):7770.
42. Neal AP, Guilarte TR. Mechanisms of lead and manganese neurotoxicity. Toxicol Res. 2013;2(2):99–114.
43. Bulcke F, Dringen R, Scheiber IF. Neurotoxicity of copper. Adv Neurobiol. 2017;18:313–43.
44. Lakshmi Priya MD, Geetha A. Level of trace elements (copper, zinc, magnesium and selenium) and toxic elements (lead and mercury) in the hair and nail of children with autism. Biol Trace Elem Res. 2011;142(2):148–58.
45. Alavanja MC. Introduction: pesticides use and exposure extensive worldwide. Rev Environ Health. 2009;24(4):303–9.
46. Boliko MC. FAO and the situation of food security and nutrition in the world. J Nutr Sci Vitaminol. 2019;65(Supplement):S4–s8.
47. Eskenazi B, Chevrier J, Rosas LG, Anderson HA, Bornman MS, Bouwman H, et al. The Pine River statement: human health consequences of DDT use. Environ Health Perspect. 2009;117(9):1359–67.
48. Eskenazi B, Marks AR, Bradman A, Fenster L, Johnson C, Barr DB, et al. In utero exposure to dichlorodiphenyltrichloroethane (DDT) and dichlorodiphenyldichloroethylene (DDE) and neurodevelopment among young Mexican American children. Pediatrics. 2006;118(1):233–41.
49. Jayaraj R, Megha P, Sreedev P. Organochlorine pesticides, their toxic effects on living organisms and their fate in the environment. Interdiscip Toxicol. 2016;9(3–4):90–100.
50. Saeedi Saravi SS, Dehpour AR. Potential role of organochlorine pesticides in the pathogenesis of neurodevelopmental, neurodegenerative, and

neurobehavioral disorders: a review. Life Sci. 2016;145:255–64.
51. Roberts EM, English PB, Grether JK, Windham GC, Somberg L, Wolff C. Maternal residence near agricultural pesticide applications and autism spectrum disorders among children in the California Central Valley. Environ Health Perspect. 2007;115(10):1482–9.
52. Richardson JR, Caudle WM, Wang M, Dean ED, Pennell KD, Miller GW. Developmental exposure to the pesticide dieldrin alters the dopamine system and increases neurotoxicity in an animal model of Parkinson's disease. FASEB J. 2006;20(10): 1695–7.
53. Dallaire R, Muckle G, Rouget F, Kadhel P, Bataille H, Guldner L, et al. Cognitive, visual, and motor development of 7-month-old Guadeloupean infants exposed to chlordecone. Environ Res. 2012;118:79–85.
54. Adeyinka A, Muco E, Regina AC, Pierre L. Organophosphates. Treasure Island: StatPearls; 2024. Disclosure: Erind Muco declares no relevant financial relationships with ineligible companies. Disclosure: Angela Regina declares no relevant financial relationships with ineligible companies. Disclosure: Louisdon Pierre declares no relevant financial relationships with ineligible companies.: StatPearls Publishing Copyright © 2024, StatPearls Publishing LLC.
55. Miodovnik A. Prenatal exposure to industrial chemicals and pesticides and effects on neurodevelopment☆. In: Nriagu J, editor. Encyclopedia of environmental health. 2nd ed. Oxford: Elsevier; 2019. p. 342–52.
56. Ross SM, McManus IC, Harrison V, Mason O. Neurobehavioral problems following low-level exposure to organophosphate pesticides: a systematic and meta-analytic review. Crit Rev Toxicol. 2013;43(1):21–44.
57. Rauh VA, Garfinkel R, Perera FP, Andrews HF, Hoepner L, Barr DB, et al. Impact of prenatal chlorpyrifos exposure on neurodevelopment in the first 3 years of life among inner-city children. Pediatrics. 2006;118(6):e1845–59.
58. Harari R, Julvez J, Murata K, Barr D, Bellinger DC, Debes F, et al. Neurobehavioral deficits and increased blood pressure in school-age children prenatally exposed to pesticides. Environ Health Perspect. 2010;118(6):890–6.
59. Sapbamrer R, Hongsibsong S. Effects of prenatal and postnatal exposure to organophosphate pesticides on child neurodevelopment in different age groups: a systematic review. Environ Sci Pollut Res Int. 2019;26(18):18267–90.
60. Guodong D, Pei W, Ying T, Jun Z, Yu G, Xiaojin W, et al. Organophosphate pesticide exposure and neurodevelopment in young Shanghai children. Environ Sci Technol. 2012;46(5):2911–7.
61. Checkoway H, Searles Nielsen S, Racette BA. Chapter 12 – environmental exposures and risks for Parkinson's disease. In: Aschner M, Costa LG, editors. Environmental factors in neurodevelopmental and neurodegenerative disorders. Boston: Academic Press; 2015. p. 253–65.
62. Sánchez-Santed F, Colomina MT, Herrero HE. Organophosphate pesticide exposure and neurodegeneration. Cortex. 2016;74:417–26.
63. Yilmaz B, Terekeci H, Sandal S, Kelestimur F. Endocrine disrupting chemicals: exposure, effects on human health, mechanism of action, models for testing and strategies for prevention. Rev Endocr Metab Disord. 2020;21(1):127–47.
64. Kajta M, Wójtowicz AK. Impact of endocrine-disrupting chemicals on neural development and the onset of neurological disorders. Pharmacol Rep. 2013;65(6):1632–9.
65. Ramírez V, Gálvez-Ontiveros Y, González-Domenech PJ, Baca M, Rodrigo L, Rivas A. Role of endocrine disrupting chemicals in children's neurodevelopment. Environ Res. 2022;203:111890.
66. Schantz SL, Widholm JJ, Rice DC. Effects of PCB exposure on neuropsychological function in children. Environ Health Perspect. 2003;111(3):357–576.
67. Schug TT, Blawas AM, Gray K, Heindel JJ, Lawler CP. Elucidating the links between endocrine disruptors and neurodevelopment. Endocrinology. 2015;156(6):1941–51.
68. Huang HB, Chen HY, Su PH, Huang PC, Sun CW, Wang CJ, et al. Fetal and childhood exposure to phthalate diesters and cognitive function in children up to 12 years of age: Taiwanese maternal and infant cohort study. PLoS One. 2015;10(6):e0131910.
69. Daniel S, Balalian AA, Insel BJ, Liu X, Whyatt RM, Calafat AM, et al. Prenatal and early childhood exposure to phthalates and childhood behavior at age 7 years. Environ Int. 2020;143:105894.
70. Huang HB, Kuo PH, Su PH, Sun CW, Chen WJ, Wang SL. Prenatal and childhood exposure to phthalate diesters and neurobehavioral development in a 15-year follow-up birth cohort study. Environ Res. 2019;172:569–77.
71. Rodríguez-Carrillo A, Mustieles V, Pérez-Lobato R, Molina-Molina JM, Reina-Pérez I, Vela-Soria F, et al. Bisphenol A and cognitive function in school-age boys: is BPA predominantly related to behavior? Neurotoxicology. 2019;74:162–71.
72. Stacy SL, Papandonatos GD, Calafat AM, Chen A, Yolton K, Lanphear BP, et al. Early life bisphenol A exposure and neurobehavior at 8years of age: identifying windows of heightened vulnerability. Environ Int. 2017;107:258–65.
73. Tewar S, Auinger P, Braun JM, Lanphear B, Yolton K, Epstein JN, et al. Association of bisphenol A exposure and attention-deficit/hyperactivity disorder in a national sample of U.S. children. Environ Res. 2016;150:112–8.
74. Perera FP, Li Z, Whyatt R, Hoepner L, Wang S, Camann D, et al. Prenatal airborne polycyclic aromatic hydrocarbon exposure and child IQ at age 5 years. Pediatrics. 2009;124(2):e195–202.

75. Lewis RC, Johns LE, Meeker JD. Serum biomarkers of exposure to perfluoroalkyl substances in relation to serum testosterone and measures of thyroid function among adults and adolescents from NHANES 2011–2012. Int J Environ Res Public Health. 2015;12(6):6098–114.
76. Oh J, Bennett DH, Calafat AM, Tancredi D, Roa DL, Schmidt RJ, et al. Prenatal exposure to per- and polyfluoroalkyl substances in association with autism spectrum disorder in the MARBLES study. Environ Int. 2021;147:106328.
77. Lelieveld J, Haines A, Burnett R, Tonne C, Klingmüller K, Münzel T, et al. Air pollution deaths attributable to fossil fuels: observational and modelling study. BMJ (Clinical research ed). 2023;383:e077784.
78. Calderón-Garcidueñas L, Gónzalez-Maciel A, Reynoso-Robles R, Delgado-Chávez R, Mukherjee PS, Kulesza RJ, et al. Hallmarks of Alzheimer disease are evolving relentlessly in metropolitan Mexico City infants, children and young adults. APOE4 carriers have higher suicide risk and higher odds of reaching NFT stage V at ≤ 40 years of age. Environ Res. 2018;164:475–87.
79. Pujol J, Martínez-Vilavella G, Macià D, Fenoll R, Alvarez-Pedrerol M, Rivas I, et al. Traffic pollution exposure is associated with altered brain connectivity in school children. NeuroImage. 2016;129:175–84.
80. Volk HE, Lurmann F, Penfold B, Hertz-Picciotto I, McConnell R. Traffic-related air pollution, particulate matter, and autism. JAMA Psychiatry. 2013;70(1):71–7.
81. Morgan ZEM, Bailey MJ, Trifonova DI, Naik NC, Patterson WB, Lurmann FW, et al. Prenatal exposure to ambient air pollution is associated with neurodevelopmental outcomes at 2 years of age. Environ Health. 2023;22(1):11.
82. Wang P, Zhao Y, Li J, Zhou Y, Luo R, Meng X, et al. Prenatal exposure to ambient fine particulate matter and early childhood neurodevelopment: a population-based birth cohort study. Sci Total Environ. 2021;785:147334.
83. Lemoine P, Harousseau H, Borteyru JP, Menuet JC. Children of alcoholic parents – observed anomalies: discussion of 127 cases. Ther Drug Monit. 2003;25(2):132–6.
84. May PA, Chambers CD, Kalberg WO, Zellner J, Feldman H, Buckley D, et al. Prevalence of fetal alcohol spectrum disorders in 4 US communities. JAMA. 2018;319(5):474–82.
85. Roozen S, Peters GJ, Kok G, Townend D, Nijhuis J, Curfs L. Worldwide prevalence of fetal alcohol spectrum disorders: a systematic literature review including meta-analysis. Alcohol Clin Exp Res. 2016;40(1):18–32.
86. Astley SJ, Aylward EH, Olson HC, Kerns K, Brooks A, Coggins TE, et al. Magnetic resonance imaging outcomes from a comprehensive magnetic resonance study of children with fetal alcohol spectrum disorders. Alcohol Clin Exp Res. 2009;33(10):1671–89.
87. Dodge NC, Thomas KGF, Meintjes EM, Molteno CD, Jacobson JL, Jacobson SW. Reduced hippocampal volumes partially mediate effects of prenatal alcohol exposure on spatial navigation on a virtual water maze task in children. Alcohol Clin Exp Res. 2020;44(4):844–55.
88. Inkelis SM, Moore EM, Bischoff-Grethe A, Riley EP. Neurodevelopment in adolescents and adults with fetal alcohol spectrum disorders (FASD): a magnetic resonance region of interest analysis. Brain Res. 2020;1732:146654.
89. Lebel C, Roussotte F, Sowell ER. Imaging the impact of prenatal alcohol exposure on the structure of the developing human brain. Neuropsychol Rev. 2011;21(2):102–18.
90. Wozniak JR, Muetzel RL. What does diffusion tensor imaging reveal about the brain and cognition in fetal alcohol spectrum disorders? Neuropsychol Rev. 2011;21(2):133–47.
91. Coles C. Critical periods for prenatal alcohol exposure: evidence from animal and human studies. Alcohol Health Res World. 1994;18(1):22–9.
92. CDC. Alcohol use and binge drinking among women of childbearing age – United States, 2006–2010. MMWR Morb Mortal Wkly Rep. 2012;61(28):534–8.
93. Hoyme HE, Kalberg WO, Elliott AJ, Blankenship J, Buckley D, Marais AS, et al. Updated clinical guidelines for diagnosing fetal alcohol spectrum disorders. Pediatrics. 2016;138(2):e20154256.
94. Mattson SN, Bernes GA, Doyle LR. Fetal alcohol spectrum disorders: a review of the neurobehavioral deficits associated with prenatal alcohol exposure. Alcohol Clin Exp Res. 2019;43(6):1046–62.
95. Wilhoit LF, Scott DA, Simecka BA. Fetal alcohol spectrum disorders: characteristics, complications, and treatment. Community Ment Health J. 2017;53(6):711–8.
96. Williams JF, Smith VC. Fetal alcohol spectrum disorders. Pediatrics. 2015;136(5):e1395–406.
97. Emberti Gialloreti L, Mazzone L, Benvenuto A, Fasano A, Alcon AG, Kraneveld A, et al. Risk and protective environmental factors associated with autism spectrum disorder: evidence-based principles and recommendations. J Clin Med. 2019; 8(2):217.
98. Zeidan J, Fombonne E, Scorah J, Ibrahim A, Durkin MS, Saxena S, et al. Global prevalence of autism: a systematic review update. Autism Res. 2022;15(5):778–90.
99. Kern JK, Geier DA, Sykes LK, Haley BE, Geier MR. The relationship between mercury and autism: a comprehensive review and discussion. J Trace Elem Med Biol. 2016;37:8–24.
100. Mold M, Umar D, King A, Exley C. Aluminium in brain tissue in autism. J Trace Elem Med Biol. 2018;46:76–82.
101. Sealey LA, Hughes BW, Sriskanda AN, Guest JR, Gibson AD, Johnson-Williams L, et al. Environmental factors in the development of

autism spectrum disorders. Environ Int. 2016;88: 288–98.
102. Yassa HA. Autism: a form of lead and mercury toxicity. Environ Toxicol Pharmacol. 2014;38(3):1016–24.
103. Panesar HK, Kennedy CL, Keil Stietz KP, Lein PJ. Polychlorinated biphenyls (PCBs): risk factors for autism spectrum disorder? Toxics. 2020; 8(3):70.
104. Huang Y, Iosif AM, Hansen RL, Schmidt RJ. Maternal polyunsaturated fatty acids and risk for autism spectrum disorder in the MARBLES high-risk study. Autism. 2020;24(5):1191–200.
105. Christensen J, Grønborg TK, Sørensen MJ, Schendel D, Parner ET, Pedersen LH, et al. Prenatal valproate exposure and risk of autism spectrum disorders and childhood autism. JAMA. 2013;309(16): 1696–703.
106. Abdelnour E, Jansen MO, Gold JA. ADHD diagnostic trends: increased recognition or overdiagnosis? Mo Med. 2022;119(5):467–73.
107. Lindblad F, Hjern A. ADHD after fetal exposure to maternal smoking. Nicotine Tob Res. 2010;12(4):408–15.
108. Pagnin D, Zamboni Grecco ML, Furtado EF. Prenatal alcohol use as a risk for attention-deficit/hyperactivity disorder. Eur Arch Psychiatry Clin Neurosci. 2019;269(6):681–7.
109. Figueroa R. Use of antidepressants during pregnancy and risk of attention-deficit/hyperactivity disorder in the offspring. J Dev Behav Pediatr. 2010;31(8):641–8.
110. Froehlich TE, Anixt JS, Loe IM, Chirdkiatgumchai V, Kuan L, Gilman RC. Update on environmental risk factors for attention-deficit/hyperactivity disorder. Curr Psychiatry Rep. 2011;13(5):333–44.

Benefits of Greenspace for Neurodevelopmental Outcomes

Mohammad Javad Zare Sakhvid, Amir Houshang Mehrparvar, and Payam Dadvand

Abstract

The rapid urbanization in recent decades has resulted in an increase in the number of children who are born and raised in urban areas. Urbanization is linked to a stressful and sedentary lifestyle, increased exposure to such environmental hazards, and limited access to natural environments, particularly greenspace. Living and rising in urban areas with less access to natural environments has prompted researchers to explore the relationship between exposure to greenspace and health outcomes, including developmental outcomes. Evidence suggests a positive role of greenspace exposure in better developmental outcomes across different domains. Additionally, evidence is accumulating on the role of greenspace in reducing the prevalence and severity of behavioral outcomes, attention deficiency, hyperactivity disorder, and autism spectrum disorder. However, the observed associations are, in some cases, not conclusive and need further studies. Moreover, the role of variables such as gender, socioeconomic status, and contextual factors in the observed associations should be explored in more detail. In terms of mechanistical pathways, several functions of greenspace—including attention restoration, mental health benefits, pollution mitigation, enriching gut microbiota, and improving prenatal mental health—could be hypothesized. Future studies need to explore the mechanistic pathways, for example, through formal mediation analyses or the use of different exposure metrics. Overall, the observed positive associations highlight the importance of community-level interventions leading to improvements in nature contact, especially for children in the rapidly urbanizing world.

Keywords

Greenspace · Nature · Neurodevelopmental disorders · Children · Mental health

M. J. Zare Sakhvid
Occupational Health Department, School of Public Health, Shahid Sadoughi University of Medical Sciences, Yazd, Iran

A. H. Mehrparvar
Industrial Diseases Research Center, Shahid Sadoughi University of Medical Sciences, Yazd, Iran

P. Dadvand (✉)
Barcelona Institute for Global Health (ISGlobal), Barcelona, Spain

CIBER Epidemiología y Salud Pública, Instituto de Salud Carlos III, Barcelona, Spain
e-mail: payam.dadvand@isglobal.org

1 Introduction

The rapid urbanization in recent decades has resulted in an increase in the number of children who are born and raised in urban areas

© The Author(s), under exclusive license to Springer Nature Singapore Pte Ltd. 2024
R. Kelishadi (ed.), *Nature, Nurture, and Neurodevelopment*,
https://doi.org/10.1007/978-981-97-7433-3_5

[1]. Living in urban areas is typically linked to a stressful and sedentary lifestyle, longer screen time, and increased exposure to environmental hazards such as air and noise pollution and heat exposure, as well as limited access to natural environments, particularly greenspace, which is more pronounced in urban areas [2]. Natural environments, including greenspace, have been suggested to enhance the health and well-being of individuals of all ages [3]. Greenspace is defined by the U.S. Environmental Protection Agency as "land that is partly or completely covered with grass, trees, shrubs, or other vegetation" [4]. Greenspace exposure has been associated with improved physical and mental health, well-being, and development in children and adolescents [5, 6].

Neurodevelopment refers to the motor, emotional, behavioral, cognitive, and social development of children [7]. Neurodevelopment is a process including complex stages that must be completed sequentially. Impairment of this sequential process could lead to neurodevelopmental disorders. The increasing trend in neurodevelopmental disorders in children has been the subject of extensive research in recent years. There are several risk factors that can cause neurodevelopmental disorders in children [8, 9]. Among the proposed risk factors, environmental exposures are one of the most important groups. Environmental exposures are often amplified in urban areas, which also have less access to natural environments. Additionally, there is socioeconomic inequality in access to the natural environment [10].These experiences can totally have a significant impact on a child's development.

Exposure to the natural environment, including greenspace, has been shown to have a positive impact on neurodevelopment in children. One of the proposed pathways for the association between exposure to greenspace and improved health and well-being, including neurodevelopment, is through the mitigation of environmental hazards (environmental pathway). Exposure to environmental hazards such as air and noise pollution and thermal stress has been reported to be associated with different neurodevelopmental disorders [11–13]. Nature, and specifically greenspace, can improve the health and developmental trajectory of children through the mitigation of environmental hazards. The ability of greenspace to provide its health benefits has also been attributed to socio-behavioral pathways, such as an increase in physical activity, social cohesion, and attention restoration [3, 6]. Other pathways, such as the role of greenspace in improving parental mental health [14], or a link between the role of nature in forming gut microbiota composition and its subsequent role in children's behavioral outcomes, have been proposed in the literature [15, 16].

Research has indicated that children and adolescents spend less time in natural environments compared to previous generations [17]. Additionally, the increasing trend of urbanization and the rise in the number of children living and rising in urban areas with less access to natural environments have prompted researchers to explore the relationship between exposure to greenspace and health outcomes in this age group. In this chapter, we have provided a summary of the current evidence on the association between greenspace exposure and various neurodevelopmental outcomes. We have attempted to present this evidence based on the type of outcomes; however, due to some extent of overlap between symptoms and diagnoses within this classification, it is possible that the presentation of evidence is mixed with other categories.

2 Cognitive Development

Cognitive development refers to the process by which a child learns and acquires skills related to thinking, learning, reasoning, problem-solving, and decision-making [18]. It is a crucial aspect of a child's development, as it affects their ability to learn, interact with others, and succeed in academic achievement and life. Cognitive development problems, on the other hand, refer to difficulties or delays in the acquisition of cognitive skills and abilities. Various factors can affect a child's cognitive development, including genetics, environmental factors, and early life experiences. The evidence on the association between natural environment exposure and cognitive development is relatively better established compared to other neurodevelopmental outcomes. A recent review of the studies on greenspace and cognitive function found that the available studies reported the associations for seven different cognitive domains (i.e., working memory capacity, processing speed, reaction and decision speed, comprehension knowledge, fluid reasoning, and visual processing); however, the majority of the studies investigated attentional functioning [19]. The review found that greenspace was equally associated with all domains of cognitive functioning. The study found no modification role of age or type of exposure assessment in the reported associations between greenspace and cognition. The review proposed further research by including state-of-the-art assessments of cognitive outcomes and diverse exposure assessment methods within both observational and experimental designs [19]. The review also highlighted the heterogeneity within exposure and outcome assessments across studies, inappropriate use of statistical methods, and "high" or "probably high" risk of bias in some studies [19].

Another review of experimental studies showed that three cognitive domains (working memory, cognitive flexibility, and attentional control) are sensitive to the restoration effect of brief exposure to greenspace (e.g., playing in a park), and a positive association between greenspace exposure and these domains could be attributed to attention restoration theory [20, 21]. Among the original studies, one of the first large-scale epidemiological studies was conducted by Dadvand et al. [22]. They used satellite data to assess the exposure of 7–10-year-old children to greenspace at home, school, and along the route between home and school. Exposure to greenspace (school greenness and a greenness index that combined greenness across residential, commuting, and school areas) was positively associated with cognitive development, defined as increases in working memory and attention abilities [22]. Overall, the available body of evidence is suggestive of the beneficial role of greenspace on cognitive development in children. However, research needs to be done on different domains and also test the sensitivity of the findings on different types of greenspace and different contexts (e.g., climatic regions).

3 Behavioral Development

Behavioral problems are common in children, with an increasing trend in societies [23]. A large number of epidemiological studies have investigated the association of exposure to nature (specifically outdoor play and greenspace) with behavioral development in children [5]. In general, the evidence on the association between exposure to greenspace and children's behavior is increasing rapidly; however, in some cases, it is not conclusive. In this section, we present the evidence on the association between nature exposure and attention deficit hyperactivity disorder (ADHD) and autism spectrum disorder (ASD).

4 Attention-Deficit Hyperactive Disorder

ADHD is a neurodevelopmental disorder of both children and adults. It is characterized by symptoms of inattention, hyperactivity, and impulsivity [24]. It is believed to be caused by a combination of genetic, environmental, and neurological factors [25, 26]. The prevalence of ADHD varies depending on the study and the population being examined, but it is estimated that approximately 5.9–7.1% of children and adolescents worldwide suffer from ADHD [27]. In recent years, there has been an increasing trend in the diagnosis of ADHD, particularly in children [28]. This may be due to greater awareness and recognition of the disorder, improved diagnostic tools, and changes in diagnostic criteria. However, some studies suggest that environmental factors such as exposure to pollutants, poor nutrition, and excessive screen time may also play a role in the increasing prevalence of ADHD [29].

The positive association of greenspace exposure with ADHD outcomes can be attributed to several pathways, including stress reduction, attention restoration, increased physical activity and social cohesion, enrichment of gut microbiota diversity, and mitigation of environmental hazards such as air and noise pollution [5]. Children with ADHD are more susceptible to stress, which can exacerbate their symptoms. Therefore, environments that prepare stress reduction conditions can be beneficial for reducing ADHD symptoms. Additionally, inattention is a key criterion for ADHD diagnosis, and research has shown that exposure to greenspace can improve attention through the attention-restoration properties of greenspace [20]. Studies have also reported a link between gut microbiota composition in infancy and subsequent behavioral outcomes [15, 16]. Children with ADHD tend to have lower gut microbial diversity, while surrounding greenspace has been found to increase microbial diversity and alter human microbiota composition [16]. However, the available findings differ according to the exposure and outcome definitions, and most are toward the positive role of greenspace exposure in preventing/ameliorating ADHD outcomes, yet a discrepancy is available in the available studies.

Numerous studies have investigated the impact of greenspace exposure on developmental outcomes in children, with a focus on the relationship between access to residential green spaces and ADHD. According to a systematic review, approximately 65% of the total analyses in the available evidence have examined the association between greenspace exposure and ADHD problems [30]. Among these analyses, approximately 55% indicated associations between greenspace exposure and a decrease in ADHD prevalence and symptoms. However, when examining the studies with a low risk of bias, the proportion of studies that found protective associations decreased to 35%. Overall, while positive associations between greenspace and ADHD have been observed in certain studies, their conclusive nature remains to be established. For example, a study conducted in New Zealand reported that the percentage of vegetation within the proximity of a child's residence was linked to a reduced risk of ADHD in boys but not girls. This association was observed only in children who had always lived in urban areas rather than in rural areas, and among children of European ancestry as opposed to those of Maori descent [31]. Another study examined the geographical distribution of ADHD in the Midwest United States and found that living further away from a park was more predictive of ADHD [32].

5 Autism Spectrum Disorder

ASD is a neurodevelopmental condition that affects communication, social interaction, and behavior [33]. While the exact cause of ASD is not fully understood, it seems to result from an interplay of genetic and environmental factors [34]. Environmental factors such as exposure to pollutants during pregnancy, infections, and birth complications have been linked to an increased risk of ASD [35]. The ASD prevalence is about 1% in children, with an increasing pattern compared to the past. It is unknown whether this increase is attributable to an increase in the risk factors of developing ASD or changes in community awareness and identification patterns [36].

With regard to mechanisms, some evidence suggests that greenspace has a direct impact on individuals with ASD by improving social cohesion and promoting mental restoration. Moreover, a number of studies have shown that pregnant women residing in areas with rich green spaces exhibit a lower level of stress [37] and prevalence of depression [14]. These mental health outcomes are particularly significant in the hypothetical pathway from greenspace to ASD, as prenatal stress, anxiety, and depression have been identified as risk factors for ASD. Thus, it is plausible that greenspace may reduce the incidence of ASD by improving parental mental health. Supporting this notion, a recent study observed a significant protective effect of prenatal exposure to greenspace on the occurrence of ASD across various buffer sizes ranging from 100 to 500 m [38]. Notably, the finding of a significant protective effect at smaller buffer sizes, such as 100 m, may suggest that greenspace imposes its benefits through mechanisms linked to visual exposure, such as mental health benefits and relaxation

[39]. Some other evidence suggests that greenspace may have an indirect role in reducing air pollution exposure and mitigating other environmental exposures that have been linked to an increased risk of ASD. For instance, one study found that a 10% increase in a forest, grassland, average tree canopy, and near-road tree canopy was associated with a decrease in autism prevalence of 10%, 10%, 11%, and 19%, respectively [40]. According to this study, the strongest inverse association was between near-road tree canopy and autism prevalence (Relative risk = 0.81, 95% CI: 0.73–0.91). However, another study found that maternal exposure to higher levels of greenspace was associated with a reduced risk of ASD in children, and this association was not mediated by air pollution, which acted more as a confounder than a mediator [38].

Recent research suggests that exposure to greenspace can have a protective impact on development and symptoms in children with ASD [38]. Despite the reported protective association between greenspace and ASD [38, 40], the causal relationship between exposure to greenspace and a lower risk of ASD is yet to be established. For example, a study using zip-code level data found higher levels of anxiety in children with ASD who were exposed to greenspace. The authors suggested that children with ASD may experience the stress-reducing effects of greenspace differently from their neurotypical peers [41]. This controversial finding has been replicated in another study, which found no association between greenspace exposure and conduct problems in children diagnosed with ASD [42].

In addition to the findings of observational studies, several experimental studies have also examined the "therapeutic effect" of greenspace or nature therapy on the improvement of ASD symptoms. A controlled study used an outdoor adventure program (30 min' sessions of challenge-based activities in urban parks for 13 weeks) to examine the role of nature in reducing ASD symptoms. The study found a significant decrease in ASD severity, including improvement in social communication and social cognition, social motivation, and subdomains of autistic mannerisms in the intervention group. The most improved subdomain was the social communication subdomain. It highlighted that greenspace can improve interpersonal relationships using adventurous activities, and through this mechanism could improve the symptoms of ASD [43]. It seems that outdoor nature-based programs can offer opportunities for children to cooperate with their peers, as they require the use of interpersonal communication to achieve group goals. Additionally, fortifying the interventions with the requirements of problem-solving skills and forcing the child to communicate in extreme, exciting, and emotionally driven situations could lead to improvement in symptoms in children with ASD [43].

The association between greenspace and ASD (and well-being in general) may depend on the health status of the population under study [42]. Therefore, generalization of the findings to a specific group should be made with caution, as the findings could be different according to different personal and contextual characteristics, such as sociodemographic and cultural factors, climate, and vegetation type. Furthermore, the timing of exposure to greenspace and other environmental factors (prenatal or childhood) may also play a role. Exploring the most influential time window of exposure also could help to understand the pathway between greenspace exposure and ASD.

6 Communication and Language Development

The acquisition of language and communication skills in the early stages of a child's development is crucial to their nonphysical growth and opportunities in life. These skills play an indispensable role in fostering educational achievement, facilitating positive social relationships, and determining future employment prospects. There is increasing evidence that demonstrates the benefits of nature for a child's language, communication, and social development [44, 45]. A qualitative study on Norwegian teenagers using thematic analysis found that the students' English fluency improved better in the outdoor environment, and this type of activity increased their confidence and intrinsic motivation for developing their spoken English skills [44]. A study examined the impact of connection to nature in 775 children in the United States and found that children who were more connected to nature had significantly higher English attainment, but not for Mathematics. Another study demonstrated specific improvements in children's language and communication skills, especially for children who did not have access to nature as part of their everyday lives, and made clear associations between outdoor nature play and children's communication and social skills [46].

In contrast to the aforementioned studies, another study found a weak beneficial association between exposure to residential greenspace and childhood development (assessed five developmental domains, including physical health, social competence, emotional maturity, cognitive and language development, and communication skills and general knowledge by teacher ratings on the Early Development Instrument [EDI]), whereas it did not observe any association between lifetime residential exposure to greenspace and scores in the language and cognitive development domains. However, the observed null association might, in part, be due to the outcome metric used in the study, which captured teacher ratings on children's skills. The study hypothesized that the weak positive association observed between greenspace exposure and early childhood development indicates that sociodemographic factors are more important in explaining variation in EDI outcomes across children than are aspects of the physical environment [47]. In addition to the influence of nature on children's communication with their peers, it could also improve the child–parent connection and communication skills. For example, a study examining whether natural environments affect communication between parents and their children found that parent–child communication is more responsive and more effectively connected in a natural environment compared to an indoor environment [45].

In addition to the large-scale observational studies on the long-term effects of greenspace on communication and language development, experimental studies have also suggested beneficial effects of brief exposure to greenspace on children with communication and language problems. An experimental study developed an ecotherapy program for a group of children with speech disturbances. All participating children in the program displayed visible signs of vocabulary enrichment, development of communication skills, increased complexity of speech structure, better psycho-emotional states, improved observation skills and self-assurance, higher creativity, and better imagination as compared to the control group [48]. However, the reported associations between nature exposure and children's communication and language development are heterogenic according to the type of exposure indicator. For example, a study in Berlin, Germany, found that the percentage of language development deficits was inversely associated with the proportion of natural area within the subdistrict of residence. On the other hand, it was associated with per capita natural area. The association with the proportion of inhabitants living no more than 300 m from a natural area was not significant [49]. All in all, the available evidence is suggestive of a beneficial association between greenspace and language and communication development; however, more studies are needed to establish this effect and replicate it in different contexts.

7 Social and Emotional Development

Social and emotional development refers to acquiring social abilities (i.e., the ability to create and sustain meaningful relationships with adults and other children) and emotional abilities (i.e., to express, recognize, and manage his or her emotions, as well as respond appropriately to others' emotions) [50]. A review of observational and experimental studies found that spending time in green spaces and greenspace availability were associated with positive adaptive behaviors in almost 22% of the analyses in the included studies, while maladaptive behaviors were inversely associated with greenspace in approximately 15% of the analyses [30]. Another review on greenspace and different behavioral outcomes found that among the studies that used the strength and difficulties questionnaire for assessing total behavioral problems, more than 80% of the reported associations were suggestive of the protective and beneficial role of greenspace exposure [5]. The findings on pro-social behaviors and emotional outcomes were suggestive of the beneficial role of greenspace exposure in more than 70% and 50% of reported associations in the retrieved studies in the review, respectively [5]. However, the relationship between greenspace and conduct problems is complex and varied based on the socioeconomic status of parents (e.g., maternal education). For example, for children with mothers in the lowest education group, living closer to urban greenspace has been reported to be associated with reduced conduct problems and increased pro-social behavior. Similarly, having access to a private garden has been reported to be associated with lower conduct problems for children from families with low levels of education. While residential vegetation has been reported to be associated with reduced conduct problems and prosocial behavior for children of highly educated mothers, no significant differences were observed for children of mothers with low levels of education [51]. Furthermore, the proportion of natural space within the child's postcode of residence, but not the proportion of park space or having access to private gardens, was reported to be positively associated with pro-social behavior for children of highly educated families but not for children from low-education households. Stratification by sex and race has not consistently revealed differential associations. A study conducted in Scotland found that conduct problems were lower among boys with access to a private garden and pro-social behavior was higher among girls with more greenspace in their postcode [52].

In summary, a growing body of evidence suggests improved social and emotional development and reduced risk of peer relationship and emotional problems in association with exposure to greenspace. Such an association could depend on various factors, such as the type of greenspace, socioeconomic status, and the child's sex and race.

8 Motor Development

Motor development refers to the transition from reflexive to voluntary and goal-directed motor actions in children. It is to demonstrate skillful, efficient, and voluntary postures and movement patterns. Motor development includes physical growth and strengthening of a child's bones, muscles, and ability to move and touch his/her surroundings [53, 54]. Motor skills are associated with a wide range of health benefits and are essential for children's physical, cognitive, and social development [55]. Neurodevelopmental motor disorders, such as developmental coordination disorder and motor delay, can significantly impact a child's daily life. It has been reported that deficits in motor development have been increasing over time [56]. Motor development disorders are considered to be multifactorial conditions, with both genetic background and environmental factors being involved in their pathogenesis [57]; however, the relative contribution of each of these etiologies is not clear. The change in the prevalence of motor development disorders over time highlights the importance of environmental impacts.

Exposure to nature has been suggested by several studies as a potential way to enhance motor development in children [30, 58]. However, direct evidence on the association between nature and motor development issues is very limited. A review found only two observational studies on this association, in which around two-thirds of the analyses suggested that greenspace was associated with better scores in fine and gross motor skills [30]. Another study assessed the association of urban greenspace with children's motor development in a group of children before enrolment in primary schools. Higher neighborhood public greenspace and composite nature in the study were significantly associated with a lower risk of motor development deficits; however, the association was not statistically significant when using the Normalized Difference Vegetation Index (NDVI) [58]. In addition to this direct evidence, indirect evidence could also be used to highlight the importance of the natural environment in neurodevelopmental motor skills. For example, available evidence on the importance of outdoor natural environments (green and blue) for improving children's health through promoting physical activity, and consequently the influence on weight status on neurodevelopmental motor skills, could be regarded as indirect evidence of the positive role of nature [58, 59].

9 Methodological Issues

9.1 Exposure Assessment

An important consideration when interpreting the findings of studies on the association between greenspace and neurodevelopment is the possibility of small direct effect sizes, which can be masked by misclassifications in exposure or small sample sizes. A review of both experimental and observational studies on the association between greenspace and blue space exposure and cognitive development found that none of the studies was rated as "low risk of bias" in the exposure assessment domain [19]. Therefore, using accurate exposure assessment methods could prevent misclassification bias and finally lead to more significant results in the studies. Additionally, given the various hypothetical pathways of greenspace on neurodevelopmental outcomes, it is recommended that future studies consider using different indicators of greenspace exposure that are pertinent to distinct proposed mechanisms. For instance, residential proximity to or the use of green spaces may be more relevant to physical activity, while mitigating urban-related hazards associated with residential or school surrounding greenspace may be more relevant to other outcomes. Existing research on greenspace and children's behavior typically employs greenspace availability indicators (surrounding greenness and percentage of land cover) as measures of exposure. However, using other categories of greenspace exposure indicators, such as greenspace accessibility, visual access to greenspace, quality of green spaces, and particularly the use of greenspace, may be more suitable to explore potential associations.

9.2 Outcome Assessment

A review on greenspace and children's behavior found that the majority of studies included in the review utilized questionnaires or assessment tools that were completed by parents, which could introduce the possibility of subjectivity bias [5]. In future studies, the use of objective measures to characterize neurodevelopmental outcomes to complement subjective measures by parents and/or teachers could help characterize the neurodevelopmental outcomes more comprehensively and robustly. Objective measures could include computerized tests or neuroimaging techniques, or the implementation of assessment tools by trained professionals such as pediatric psychologists or psychiatrists. Additionally, the use of multiple sources for questionnaires, such as gathering information from both the child and their parents or teachers, could help improve the validity and reliability of outcomes. It should be noted that using different outcome assessment methods, such as self-administered questionnaires, direct interviews, or diagnosis and evaluation by healthcare professionals, can also introduce heterogeneity in the findings due to differences in their validity and reliability in characterizing neurodevelopmental problems. However, these different tools can also provide valuable information on different aspects of neurodevelopmental problems, which may complement each other to provide a more robust assessment of outcomes. Therefore, incorporating a variety of assessment methods may improve the overall characterization of behavioral problems in future studies.

9.3 Role of Confounders and Modifiers

The effect of the natural environment on health could be modified or confounded by factors such as socioeconomic status (SES) (e.g., parental education), sex, ethnicity, and the degree of urbanity [5]. Special care should be taken with SES variables. In addition to being a potential effect modifier for the association between greenspace exposure and neurodevelopment, SES can be an influential confounder as well. As such, studies evaluating this association need to properly control their analyses for SES indicators at both household (e.g., parental education, income, and employment status) and neighborhood (e.g., census-based indices of deprivation) levels to minimize the likelihood of the residual SES confounder. The included articles in a recent review on greenspace and children's behavior found that the analyses were mainly controlled for indicators of household SES without taking neighborhood SES into account, which remains an area for improvement in future studies [5].

The role of SES in the association between nature exposure and developmental outcomes could be more complicated and also outcome-dependent. Previous research has demonstrated that individuals with lower socioeconomic status (measured by various indicators such as income, education, and occupation) typically reside in areas with lower availability of greenspace. Furthermore, a higher prevalence of some developmental outcomes, such as ASD, has been reported in families with higher socioeconomic status, such as those with greater education [60]. Conversely, parents with lower socioeconomic status may have limited access to healthcare services to diagnose or treat their children with ASD. These factors could result in an underestimation of the risk of ASD in areas with limited greenspace, particularly among those with lower socioeconomic status. Consequently, the inclusion of socioeconomic status variables in greenspace studies is essential, and future studies need to control for wider SES variables.

9.4 Intermediate Factors

The use of mediation analysis is a formal way of exploring pathways in observational epidemiological studies. However, most of the studies have not explored the mechanisms through the mediation analyses. A study conducted in Canada examined the association between lifelong exposure to residential greenspace and early childhood development while assessing the degree to which this observed association is influenced by the inclusion of traffic-related air pollution and noise. The findings indicated that a rise in the proportion of vegetation was associated with minor advancements in early childhood development. Further analyses revealed that the inclusion of NO_2, $PM_{2.5}$, and noise in the models mediated 97.1% (95% CI 43.0–396.0), 29.5% (12.0–117.0), and 35.2% (17.9–139.0) of this association, respectively. Consequently, it can be inferred that increased exposure to residential greenspace could enhance early childhood development by mitigating the negative developmental effects of traffic-related exposures, particularly NO_2 [47]. Another study on developmental motor deficits found that, following the adjustment of models for air pollution and noise, the effect sizes attenuated. This evidence also highlights the potential mediating effect of air and noise pollution [58]. Similarly, a study on the association between exposure to greenspace at home, school, and commuting in school children found that the addition of traffic-related air pollutants to models explained 20–65% of the estimated associations between school greenness and 12-month cognitive development [22].

Most of the understanding about possible mediators in the association between nature exposure and developmental outcomes comes from studies that used the "additional adjustment with possible mediators" approach. The suitability of this approach is questionable, and the results could be misleading. We recommend that future studies formally employ mediation analysis to explore the mediating role of air pollution, physical activity, obesity, and environmental microbiota in the association between greenspace exposure and behavior problems.

10 Further Research

The available literature on the association between exposure to greenspace and children's neurodevelopment primarily originates from North American and European countries. Unfortunately, low- and middle-income countries, particularly those located in Africa, South America, and the Middle-East, are notably underrepresented in the current literature. It is necessary to investigate the role of context in the association between nature exposure and children's development.

To establish the causal nature of the associations observed in previous studies, future research should systematically evaluate the potential mechanisms that underlie these findings. Using different outcome assessment tools, including both objective and subjective measures from different sources (e.g., parents, teachers, and healthcare professionals), could increase the likelihood of detecting the often modest effects of greenspace on neurodevelopmental outcomes [19]. Furthermore, the limited number of studies with comparable exposure and outcome metrics presents a significant obstacle when comparing and pooling studies in meta-analyses for summarizing the effect size of the associations. As a result, researchers are limited in conducting meta-analyses to identify outcome-specific sources of heterogeneity and compare these sources among different outcomes. To overcome this limitation, it is recommended that future studies use standardized and appropriate exposure indicators and outcome instruments to enhance comparability between studies.

11 Conclusion

The observed positive association between nature exposure and better neurodevelopmental outcomes is promising because environmental factors are more modifiable than genetic factors. It indicates that incorporating nature exposure into the daily lives of children could potentially prevent unwanted developmental outcomes. Additionally, despite accumulating evidence on the beneficial association between the natural environment and children's development, we still see a decreasing trend in time spent on outdoor activities by children across the globe [61]. This highlights the need for interventions leading to improvement in nature contact, especially for children in our rapidly urbanizing world.

References

1. United Nations-Population Division, United Nations. World urbanization prospects: The 2014 revision, highlights. Department of Economic and Social Affairs. 2014.
2. Vlahov D. Urbanization, urbanicity, and health. J Urban Health. 2002;79(90001):1S–2.
3. Markevych I, Schoierer J, Hartig T, Chudnovsky A, Hystad P, Dzhambov AM, et al. Exploring pathways linking greenspace to health: theoretical and methodological guidance. Environ Res. 2017;158:301–17.
4. US E.P.A. What is open space/green space [internet]. 2021. Available from: https://www3.epa.gov/region1/eco/uep/openspace.html
5. Sakhvidi MJZ, Knobel P, Bauwelinck M, de Keijzer C, Boll LM, Spano G, et al. Greenspace exposure and children behavior: a systematic review. Sci Total Environ. 2022;153608:153608.
6. Zare Sakhvidi MJ, Mehrparvar AH, Zare Sakhvidi F, Dadvand P. Greenspace and health, wellbeing, physical activity, and development in children and adolescents: an overview of the systematic reviews. Curr Opin Environ Sci Health. 2023;32:100445.
7. Laura E, Levine EM. Child development; an active learning approach. 2014.
8. Vilcins D, Sly PD, Jagals P. Environmental risk factors associated with child stunting: a systematic review of the literature. Ann Glob Health. 2018;84(4):551–62.
9. Bignardi G, Dalmaijer ES, Astle DE. Testing the specificity of environmental risk factors for developmental outcomes. Child Develop [Internet]. 2022;93(3) [cited 2023 Apr 3]. Available from: https://onlinelibrary.wiley.com/doi/10.1111/cdev.13719
10. Strife S, Downey L. Childhood development and access to nature: a new direction for environmental inequality research. Organ Environ. 2009;22(1):99–122.
11. Costa LG, Cole TB, Dao K, Chang YC, Coburn J, Garrick JM. Effects of air pollution on the nervous system and its possible role in neurodevelopmental and neurodegenerative disorders. Pharmacol Ther. 2020;107523:107523.
12. Zare Sakhvidi F, Zare Sakhvidi MJ, Mehrparvar AH, Dzhambov AM. Environmental noise exposure and neurodevelopmental and mental health problems in children: a systematic review. Curr Envir Health Rpt. 2018;5(3):365–74.
13. Yu T, Zhou L, Xu J, Kan H, Chen R, Chen S, et al. Effects of prenatal exposures to air sulfur dioxide/nitrogen dioxide on toddler neurodevelopment and effect modification by ambient temperature. Ecotoxicol Environ Saf. 2022;230:113118.
14. McEachan RRC, Prady SL, Smith G, Fairley L, Cabieses B, Gidlow C, et al. The association between green space and depressive symptoms in pregnant women: moderating roles of socioeconomic status and physical activity. J Epidemiol Community Health. 2016 Mar;70(3):253–9.

15. Loughman A, Ponsonby AL, O'Hely M, Symeonides C, Collier F, Tang MLK, et al. Gut microbiota composition during infancy and subsequent behavioural outcomes. EBioMedicine. 2020;52:102640.
16. Kalenik A, Kardaś K, Rahnama A, Sirojć K, Wolańczyk T. Gut microbiota and probiotic therapy in ADHD: a review of current knowledge. Prog Neuro-Psychopharmacol Biol Psychiatry. 2021;110:110277.
17. Gaster S. Urban children's access to their neighborhood: changes over three generations. Environ Behav. 1991;23(1):70–85.
18. Gauvain M, Richert R. Cognitive development. In: Reference module in neuroscience and biobehavioral psychology [Internet]. Elsevier; 2022. p. B9780323914970001000. [cited 2023 Apr 5]. Available from: https://linkinghub.elsevier.com/retrieve/pii/B9780323914970000485
19. Buczyłowska D, Zhao T, Singh N, Jurczak A, Siry A, Markevych I. Exposure to greenspace and bluespace and cognitive functioning in children—a systematic review. Environ Res. 2023;222:115340.
20. Ohly H, White MP, Wheeler BW, Bethel A, Ukoumunne OC, Nikolaou V, et al. Attention restoration theory: a systematic review of the attention restoration potential of exposure to natural environments. J Toxic Environ Health, Part B. 2016;19(7):305–43.
21. Stevenson MP, Schilhab T, Bentsen P. Attention Restoration Theory II: a systematic review to clarify attention processes affected by exposure to natural environments. J Toxic Environ Health, Part B. 2018;21(4):227–68.
22. Dadvand P, Nieuwenhuijsen MJ, Esnaola M, Forns J, Basagaña X, Alvarez-Pedrerol M, et al. Green spaces and cognitive development in primary schoolchildren. Proc Natl Acad Sci. 2015;112(26):7937–42.
23. Tick NT, van der Ende J, Verhulst FC. Twenty-year trends in emotional and behavioral problems in Dutch children in a changing society. Acta Psychiatr Scand. 2007;116(5):473–82.
24. Fitzgerald M, Bellgrove M, Gill M. Handbook of attention deficit hyperactivity disorder. 2007.
25. Sciberras E, Mulraney M, Silva D, Coghill D. Prenatal risk factors and the etiology of ADHD—review of existing evidence. Curr Psychiatry Rep. 2017 Jan;19(1):1.
26. Kim JH, Kim JY, Lee J, Jeong GH, Lee E, Lee S, et al. Environmental risk factors, protective factors, and peripheral biomarkers for ADHD: an umbrella review. Lancet Psychiatry. 2020;7(11):955–70.
27. Willcutt EG. The prevalence of DSM-IV attention-deficit/hyperactivity disorder: a meta-analytic review. Neurotherapeutics. 2012;9(3):490–9.
28. Davidovitch M, Koren G, Fund N, Shrem M, Porath A. Challenges in defining the rates of ADHD diagnosis and treatment: trends over the last decade. BMC Pediatr. 2017 Dec;17(1):218.
29. Banerjee TD, Middleton F, Faraone SV. Environmental risk factors for attention-deficit hyperactivity disorder. Acta Paediatr. 2007;96(9):1269–74.
30. Mygind L, Kurtzhals M, Nowell C, Melby PS, Stevenson MP, Nieuwenhuijsen M, et al. Landscapes of becoming social: a systematic review of evidence for associations and pathways between interactions with nature and socioemotional development in children. Environ Int. 2021;146:106238.
31. Donovan GH, Michael YL, Gatziolis D, Mannetje A, Douwes J. Association between exposure to the natural environment, rurality, and attention-deficit hyperactivity disorder in children in New Zealand: a linkage study. Lancet Planet Health. 2019;3(5):e226–34.
32. Baumgardner DJ, Schreiber AL, Havlena JA, Bridgewater FD, Steber DL, Lemke MA. Geographic analysis of diagnosis of attention-deficit/hyperactivity disorder in children: eastern Wisconsin, USA. Int J Psychiatry Med. 2010;40(4):363–82.
33. Lord C, Elsabbagh M, Baird G, Veenstra-Vanderweele J. Autism spectrum disorder. Lancet. 2018;392(10146):508–20.
34. Kim YS, Leventhal BL. Genetic epidemiology and insights into interactive genetic and environmental effects in autism spectrum disorders. Biol Psychiatry. 2015;77(1):66–74.
35. Sealey LA, Hughes BW, Sriskanda AN, Guest JR, Gibson AD, Johnson-Williams L, et al. Environmental factors in the development of autism spectrum disorders. Environ Int. 2016;88:288–98.
36. Rice CE, Rosanoff M, Dawson G, Durkin MS, Croen LA, Singer A, et al. Evaluating changes in the prevalence of the Autism Spectrum Disorders (ASDs). Public Health Rev. 2012;34(2):17.
37. Barry F, Riis V, Ncube L, Elovitz MA, South EC, Burris HH. Self-reported greenspace exposure and perceived stress in a Black pregnancy cohort. Am J Obstet Gynecol. 2023;228(1):S484–5.
38. Pagalan L, Oberlander TF, Hanley GE, Rosella LC, Bickford C, Weikum W, et al. The association between prenatal greenspace exposure and Autism spectrum disorder, and the potentially mediating role of air pollution reduction: a population-based birth cohort study. Environ Int. 2022;167:107445.
39. Labib SM, Lindley S, Huck JJ. Scale effects in remotely sensed greenspace metrics and how to mitigate them for environmental health exposure assessment. Comput Environ Urban Syst. 2020;82:101501.
40. Wu J, Jackson L. Inverse relationship between urban green space and childhood autism in California elementary school districts. Environ Int. 2017;107: 140–6.
41. Larson LR, Barger B, Ogletree S, Torquati J, Rosenberg S, Gaither CJ, et al. Gray space and green space proximity associated with higher anxiety in youth with autism. Health Place. 2018;53:94–102.

42. Barger B, Larson LR, Torquati J, Moody E, Ogletree S, Rosenberg S. The complex relationship between greenspace and well-being in children with and without autism. Appl Psych Health Well. 2022;aphw.12406
43. Zachor DA, Vardi S, Baron-Eitan S, Brodai-Meir I, Ginossar N, Ben-Itzchak E. The effectiveness of an outdoor adventure programme for young children with autism spectrum disorder: a controlled study. Dev Med Child Neurol. 2017;59(5):550–6.
44. Myhre TS, Fiskum TA. Norwegian teenagers' experiences of developing second language fluency in an outdoor context. J Adventure Edu Out Learn. 2021;21(3):201–16.
45. Cameron-Faulkner T, Melville J, Gattis M. Responding to nature: natural environments improve parent-child communication. J Environ Psychol. 2018;59:9–15.
46. O'Brien L. Learn Out: the Forest School Approach. Education 3–13. 2009;37(1):45–60.
47. Jarvis I, Davis Z, Sbihi H, Brauer M, Czekajlo A, Davies HW, et al. Assessing the association between lifetime exposure to greenspace and early childhood development and the mediation effects of air pollution and noise in Canada: a population-based birth cohort study. Lancet Planetary Health. 2021;5(10): e709–17.
48. Kalashnikova IV, Gontar OB, Zhirov VK, Kalashnikov AO. Integrated animal-assisted and plant-assisted ecotherapy for preschool children with speech disturbances: a program for the Arctic. Ecopsychology. 2016;8(2):79–88.
49. Kabisch N, Haase D, Annerstedt van den Bosch M. Adding natural areas to social indicators of intra-urban health inequalities among children: a case study from Berlin, Germany. IJERPH. 2016 Aug 4;13(8):783.
50. U.S. Department of Health and Human Services. Social and Emotional Development [Internet]. U.S. Department of Health and Human Services; Available from: https://eclkc.ohs.acf.hhs.gov/school-readiness/effective-practice-guides/social-emotional-development
51. Balseviciene B, Sinkariova L, Grazuleviciene R, Andrusaityte S, Uzdanaviciute I, Dedele A, et al. Impact of residential greenness on preschool children's emotional and behavioral problems. Int J Environ Res Public Health. 2014;11(7): 6757–70.
52. Richardson EA, Pearce J, Shortt NK, Mitchell R. The role of public and private natural space in children's social, emotional and behavioural development in Scotland: a longitudinal study. Environ Res. 2017;158:729–36.
53. Hallemans A, Verbeque E, Van de Walle P. Motor functions. In: Handbook of clinical neurology [Internet]. Elsevier; 2020. p. 157–70. [cited 2023 Apr 10]. Available from: https://linkinghub.elsevier.com/retrieve/pii/B9780444641502000150
54. von Hofsten C. Motor and physical development: manual. In: Encyclopedia of infant and early childhood development [Internet]. Elsevier; 2008. p. 374–84. [cited 2023 Apr 10]. Available from: https://linkinghub.elsevier.com/retrieve/pii/B9780123708779001055
55. Lubans DR, Morgan PJ, Cliff DP, Barnett LM, Okely AD. Fundamental movement skills in children and adolescents: review of associated health benefits. Sports Med. 2010;40(12):1019–35.
56. Fleuren KMW, Smit LS, Stijnen T, Hartman A. New reference values for the Alberta Infant Motor Scale need to be established. Acta Paediatr. 2007;96(3):424–7.
57. Pennington B. From single to multiple deficit models of developmental disorders. Cognition. 2006;101(2):385–413.
58. Kabisch N, Alonso L, Dadvand P, van den Bosch M. Urban natural environments and motor development in early life. Environ Res. 2019;179(Pt A):108774.
59. Timmons BW, LeBlanc AG, Carson V, Connor Gorber S, Dillman C, Janssen I, et al. Systematic review of physical activity and health in the early years (aged 0–4 years). Appl Physiol Nutr Metab. 2012;37(4):773–92.
60. King MD, Bearman PS. Socioeconomic status and the increased prevalence of autism in California. Am Sociol Rev. 2011;76(2):320–46.
61. Scott S, Gray T, Charlton J, Millard S. The impact of time spent in natural outdoor spaces on children's language, communication and social skills: a systematic review protocol. IJERPH. 2022;19(19):12038.

Maternal Nutrition and Child Neurodevelopment

Maryam Aref, Bahar Golastani, Motahar Heidari-Beni, Maryam Bemanalizadeh, and Roya Kelishadi

Abstract

Early life environment is associated with offspring neurodevelopment. Maternal under- or over-nutrition and food choices may influence fetal neurodevelopment directly or indirectly. The role of maternal nutritional factors in fetal development was one of the important research topics during the twentieth century. During fetal neurodevelopment, adequate micronutrient and macronutrient intake is required. Thus, a fetus may be more susceptible to nutrient deficiencies. Prenatal nutrition may change developmental processes including synaptic plasticity, neurogenesis, and dendritic arborization.

Over-nutrition, such as high-fat and carbohydrate diets, is correlated with inflammatory processes that can be detrimental to offspring brain development.

Current evidence has demonstrated that single-nutrient supplementation is less adequate than supplementation with more complex formulae. However, the determination of the optimal content of micronutrient supplements and their long-term impacts on child neurodevelopment requires more investigation.

The objective of the present chapter is to provide a background and update on the current knowledge related to maternal nutrition and child neurodevelopment.

Keywords

Maternal nutritional physiological phenomena · Nutritional status · Micronutrients · Nutrients · Child development

1 Introduction

Neurodevelopment of the fetus begins approximately 22 days after conception and develops rapidly in the second and third trimesters [1, 2]. Many factors, including maternal nutrition during pregnancy or lactation, the mother's lifestyle [3], environmental pollution [4], and food contamination with chemical compounds [5] affect children's neurodevelopment.

The potential mechanisms for explaining the effects of mothers' nutrition on neurodevelopment in children are epigenetics of the fetal

M. Aref · B. Golastani · M. Bemanalizadeh · R. Kelishadi
Child Growth and Development Research Center, Research Institute for Primordial Prevention of Non-Communicable Disease, Isfahan University of Medical Sciences, Isfahan, Iran

M. Heidari-Beni (✉)
Department of Nutrition, Child Growth and Development Research Center, Research Institute for Primordial Prevention of Non-Communicable Disease, Isfahan University of Medical Sciences, Isfahan, Iran
e-mail: motahar.heidari@nutr.mui.ac.ir

© The Author(s), under exclusive license to Springer Nature Singapore Pte Ltd. 2024

R. Kelishadi (ed.), *Nature, Nurture, and Neurodevelopment*,
https://doi.org/10.1007/978-981-97-7433-3_6

genome and the regulation of fetal brain-derived neurotrophic factor (BDNF) [6].

Different nutrients affect neurodevelopment by different mechanisms. Omega-3 polyunsaturated fatty acid (ω-3 PUFA) may play critical roles in neural development. In addition, several pieces of evidence showed that higher ω-3 PUFA levels are associated with a lower risk of attention deficit hyperactivity disorder (ADHD) [7, 8]. Another important element in neurodevelopment in children is iodine. A study among 6–7-year-old children showed a positive correlation between urinary iodine concentration and cognitive function [9]. Findings regarding the role of vitamin D in child growth and development are conflicting [10–12]. A cohort study among 141 mother–infant dyads demonstrated that infants whose mothers had vitamin D deficiency during pregnancy finally reached lower ages and stages questionnaire (ASQ) scores, in which gross motor function was significantly impaired at 3 months of age [11]. On the other hand, another cohort study reported no significant associations between maternal 25(OH)-vitamin D concentrations and the child's anthropometric indices or child's intelligence outcomes [12]. A recent meta-analysis supports that increased prenatal concentration of 25(OH)D is associated with improved cognitive development and reduced risk of ADHD and autism-related traits later in life [13]. Children would have better neurodevelopment when their mothers consume sufficient fresh vegetables because fresh vegetables are rich in folate, vitamin C, beta-carotene (provitamin A), and other nutrients [6, 14].

Maternal obesity is another important factor that is involved in the neurodevelopment of the fetus because chronic inflammation can occur more frequently in obese women [15]. The placenta can mediate these inflammatory changes [16]; therefore, the presence of obesity in pregnant women can induce inflammation in the fetus. As a result, some problems occur related to brain development in the fetus [17]. Moreover, preclinical results are in accordance with these findings, showing that increasing inflammation levels can follow maternal obesity in the neonatal period and late childhood. Inflammation affects brain development both directly and indirectly. It could be toxic to brain cell development, especially oligodendrocytes [18].

A cohort study demonstrates that children of mothers with higher BMI scores showed lower cognitive ability, motor function, and academic performance [19].

Both lower weight gain and excessive weight gain during pregnancy would have their own adverse effects. There is a negative association between inadequate maternal weight gain and language skills and academic achievement, whereas having excess weight was associated with a lower Mental Developmental Index score (using the Bayley Scale of Infants and Toddlers' Development) at 2 years [20]. Preclinical models confirm that calorie restriction is involved in the neuronal system and its function. Epigenetic modifications are another responsible mechanism that can follow maternal obesity because of the inflammatory response [21]. Maternal inflammation is active in utero through cytokines and chemokines, leading to changes in cell development followed by DNA methylation changes [22].

Several studies have shown that consuming sufficient amounts of vitamins, minerals, essential elements, protein, and fatty acids during pregnancy plays a crucial role in promoting child brain development [23]. The results showed that pregnant women with a better diet and rich in nutrients exhibit better neurodevelopmental effects in their children. Fiber can also influence neurodevelopment indirectly by changing gut microbiome function and making the neurodevelopment process easier [23].

Calcium is another important element that contributes to neurodevelopment by reducing blood lead [24], facilitating the formation of neural synaptic connections and neurotransmission [25, 26], and improving thyroid function [27].

Consumption of higher levels of saturated fat during pregnancy would have negative influences on fetal brain development. On the other hand, some literature also indicates that medium-chain triglycerides, which are found high in saturated fat of certain foods, would be beneficial for neurodevelopment. However, more evidence is required to consider saturated fats as an essential factor in neurodevelopment [23].

Providing sufficient quantity and high-quality nutrients during pregnancy will affect both the mother's health and the offspring's development [28, 29].

The epigenetic profile in offspring is dependent on the mother's dietary intake. A genome-wide association study found that mothers should have a healthy and balanced nutrition intake [30]. Both genetic factors and environmental factors might lead to neurodevelopmental disorders. Moreover, maternal diet can affect genetic vulnerability, fetal genetics, and inflammation in both mother and fetus, which can influence neurodevelopment.

In the present chapter, we summarized the association between maternal nutrition during pregnancy and neurodevelopment in offspring.

2 Role of Folate on Child Neurodevelopment

Folate plays a critical role in early life. Pregnant women are advised to take folate (folic acid) from before conceiving until the end of the first trimester due to its effect on the onset and progression of the disease of neural tube defects (NTD). New evidence has shown that there is an association between maternal folate concentration and the development of myelination in the fetal brain [31] and cognitive development [32]. However, it needs more investigation. Children of mothers who had folate-related anemia in the third trimester of pregnancy suffered more from intellectual disability compared to those who had optimal folate levels.

Results showed that 11-year-old children whose mothers took folic acid during the second and third trimesters of pregnancy had significantly higher cognitive performance compared to children whose mothers took a placebo [33].

Although more studies are required to understand the precise mechanism of folic acid on child neurodevelopment, folic acid plays a role in some complex pathways that are involved in myelination, neurotransmitter synthesis, and methylation [34].

Another possible mechanism is related to DNA methylation [35]. The results of an epigenetic study suggested that taking folic acid during pregnancy had a significant influence on the methylation of specific genes, including insulin-like growth factor 2 (IGF2), brain-derived neurotrophic factor (BDNF), and long interspersed element 1 (LINE-1), which are involved in brain development [36, 37]. Although previous findings suggest mothers take folic acid from before conceiving until the end of the third trimester of pregnancy, recent evidence recommends continuing folic acid longer than the previously recommended period due to its beneficial effects on neurocognitive development in offspring [33].

3 Role of Iron on Child Neurodevelopment

The most prevalent deficiency among pregnant women, infants, and young children is iron deficiency [38]. Iron deficiency affects gene regulation following signaling cascades through the mechanistic target of rapamycin (mTOR), BDNF, and microtubule-associated protein 2 (MAP 2), which may have long-term consequences [39]. Iron is also required for producing enzymes such as dopamine, adrenaline, noradrenaline, and serotonin. Thus, iron deficiency affects memory, emotion, and executive function [40].

Studies showed that infants with iron deficiency in prenatal (cord ferritin <75 μg/L or ZPP/Heme >118 μmol/mol) or postnatal periods (body iron <0 mg/kg) had worse performance in gross motor tasks and lower memory response compared to iron-sufficient infants [41–43].

Four studies showed a positive correlation between Iron levels and infant responses [44, 45]. However, some studies showed no association between iron status and motor development. For example, a study on 828 participants in Benin and 523 participants in Vietnam did not identify any association between maternal Iron levels and motor development at 12 months and 6 months, respectively [46, 47]. Several studies that were conducted in populations from Bangladesh, France, and the United States did not report any differences in cognitive outcomes among iron-deficient or iron-sufficient participants [40].

A randomized controlled trial study in Australia showed that there was no difference in cognitive scores between groups that took 20 mg/d of iron and a placebo [48]. The lack of association between iron status and cognitive development may be related to the consideration of several aspects of cognition, including memory, executive function, attention, and language skills, which all report as a single total score [43, 49]. Studies that used specific tests could show a significant association between iron status and cognition [50].

Some evidence focuses on the importance of timing in using iron supplementation, showing that taking supplementation in early infancy (0–6 months) is more beneficial for motor development compared to using it later in infancy [40].

Several long-term consequences of iron deficiency have been reported, including delayed neural response [51, 52] and physical reaction times [51] at 10 years of age, higher rates of internalizing and externalizing behavioral problems in adolescence [53], consistently lower scores in intelligence assessments up to 19 years of age [54], reduced neural connectivity at 22 years of age [55], and poorer educational achievement, interpersonal relationships, and mental health at 25 years of age [56].

4 Role of Selenium on Child Neurodevelopment

Selenium is an essential trace element that is involved in various biological processes, including neurodevelopment. Selenium is a cofactor for glutathione peroxidase (also known as selenoprotein). Both low levels and excess levels of selenium might have adverse effects; for example, excess selenium can cause neurologic changes, increased risk of type 2 diabetes, and damage to the gastrointestinal tract [57].

There is an association between maternal selenium levels and neurodevelopment disorders in offspring, including autism spectrum disorder (ASD) and attention deficit hyperactivity disorder (ADHD), because selenium can cross the placenta [58].

A study among 1500 mother–child showed that babies whose mothers had a higher level of selenium in their blood would be at higher risk of getting autism spectrum disorder (ASD) or attention deficit hyperactivity disorder (ADHD). The underlying mechanisms of this association require further studies [59].

However, increasing selenium levels to toxic levels leads to a decrease in the number of neuronal synapses and an increase in cell death in rat hippocampal [60]. In addition, maternal folate and vitamin B12 levels are considered as other involved factors that are associated with ASD. In a prospective birth cohort study, a significant dose-dependent association between maternal RBC selenium levels and child risk of developing ASD or ADHD was found. 60 μg of selenium is recommended daily for pregnant women. The tolerable upper level is 400 μg/day for pregnant women [59].

5 Role of Vitamin B12 on Child Neurodevelopment

Vitamin B12 (cobalamin) is a water-soluble vitamin that is synthesized by microorganisms [61]. The human body cannot produce vitamin B12; thus, humans should take supplements or consume foods. The prevalence of vitamin B12 deficiency is high in several low- and middle-income countries [62].

Vitamin B12 is involved in the synthesis of deoxyribonucleic acid (DNA) and ribonucleic acid (RNA), lipids, and proteins. It plays a critical role in the development, myelination, and normal function of the central and peripheral nervous systems [63].

Vitamin B12 levels in pregnant women are important because its deficiency may lead to complications, including neural tube defects, spontaneous abortion, premature birth, and possibly low birth weight [64, 65]. Vitamin B12 status in pregnant women will affect cobalamin status in the neonate [66]. However, clinical trials showed inconsistent results regarding the association between maternal vitamin B12 deficiency and the infant's neurodevelopment [63].

Vitamin B12 deficiency can appear mostly at the age of 4–6 months through clinical symptoms, including developmental delay, muscular hypotonia, irritability, regression, feeding difficulties, and failure to thrive, as well as tremors and seizures [63, 67]. Recent evidence recommends a serum vitamin B12 level of more than 394 pmol/L at week 18 of pregnancy, and pregnant women should pay attention to their diet in order to prevent future consequences in neonatal [66, 68]. Early detection of maternal vitamin B12 deficiency is very important during pregnancy. However, the detection of vitamin B12 deficiency would be complicated because maternal blood counts alone would not be sufficient to differentiate vitamin B12 deficiency from other deficiencies [63].

6 Role of Choline on Child Neurodevelopment

The first 1000 days of life (2 years) play a significant role in appropriate neurodevelopment and brain function. Thus, taking sufficient nutrients is essential for appropriate neurodevelopmental function.

Choline has recently been considered a "neuroprotectant" and "neurocognitive essential nutrient." It is important for normal growth and brain function in early life [69, 70]. Acetylcholine, phospholipid, and betaine synthesis require choline. The hepatic phosphatidylethanolamine N-methyltransferase (PEMT) pathway can provide a few amounts of necessary choline in the human body; therefore, choline should be provided through dietary sources [69].

Choline is essential in metabolic compounds such as phospholipid and betaine synthesis. Choline is required for several processes in neural progenitor cells in the brain, including differentiation, migration, proliferation, apoptosis, and methylation; thus, choline can regulate gene expression and protein synthesis, which are involved in memory and learning [71].

Some evidence is consistent about the association between using maternal choline supplementation and the infants' intelligence. A review with 22 included studies showed that taking choline supplements by pregnant women may help improve the behavioral, cognitive, and neurological deficits in infants [72]. A US cohort suggested that getting a higher dose of choline was associated with the child's visual memory at age 7 [73]. Three other cohort studies found a positive correlation between infant neurodevelopment and choline status during pregnancy [74–76].

One study did not find any significant association between maternal choline consumption in the first or second trimesters and childhood cognition at 3 years of age. Providing the necessary amount of choline for the body depends on the amount of proper food intake and regular dietary intake [76, 77].

Several studies showed that choline levels in pregnant women affected infant processing speed and visuospatial memory in childhood [78, 79].

Using prenatal choline supplementation from 27 weeks gestation until delivery leads to better color-location memory in infants [78]. Moreover, supplement use (100 and 550 mg of choline/day as supplementation) with dietary intake (380 mg of choline/day) from the third trimester of gestation will enhance processing speed in infants [79]. One study demonstrated that consuming 900 mg of choline/day from the second trimester negatively correlated with attention problems in 40-month-old children [80]. In addition, some studies indicated a negative association between taking choline supplementation and the risk of neural tube defects (NTDs) [71].

Several animal studies suggested a protective role of choline in individuals with Down syndrome due to its potential to protect the basal forebrain and improve brain function; however, more human studies are required in this field [81–83]. Breast milk is choline-rich; thus, lactating women require a higher level of choline compared to nonlactating women to increase choline concentration in breast milk. A daily intake of 450–1000 mg of choline is recommended as an optimal dose for pregnant women in order to have a healthy infant in terms of brain function and neurodevelopment [71].

7 Role of Iodine on Child Neurodevelopment

Iodine is an essential thyroid hormone component that is involved in neurogenesis, neuronal migration, axon and dendrite growth, synaptogenesis, and myelination; therefore, normal brain growth and cognitive development need adequate iodine intake, and iodine deficiency can cause intellectual disability [84].

World Health Organization states that pregnant women, lactating mothers, and infants who live in countries with iodized salt programs do not require iodine supplementation [85]. However, the American Thyroid Association, the Endocrine Society, and the American Academy of Pediatrics recommend that pregnant women and lactating mothers should take iodine supplements daily (containing 150 µg of iodine) [86]. More studies are needed to determine how iodine affects brain development in children. However, infants who feed on breast milk should be provided with sufficient iodine to achieve optimal brain growth. Higher breast milk iodine concentration is related to better infant motor development [87].

Some studies focus on the timing of iodine supplement usage. Taking iodine supplements is requisite as early as possible during pregnancy [88]. The results of a randomized clinical trial showed that consuming 150 µg/day of iodine during lactation led to a higher level of cognitive development in children. Taking iodine supplements at doses higher than recommended does not lead to improvements or delays in neurodevelopmental outcomes [84].

Iodine deficiency in pregnant women causes adverse neurodevelopment, including cretinism and mental retardation in infants; therefore, it is highly recommended that pregnant women use iodine supplementation. A UK study and an Australian study found that lower levels of intelligence in children aged 8–9 years were correlated with deficient iodine status (urinary iodine concentration (UIC) of <150 µg/L or 150 µg/g creatinine [2]). However, a cohort study found no association between iodine deficiency in mothers during pregnancy and the intelligence quotient (IQ) score of the children [89].

Bayley Scales of Infant and Toddler Development, Third Edition (Bayley-III) is the common measurement of neurodevelopment in children [90]. Not only lower levels of iodine but also higher levels of iodine intake in pregnant women might lead to poor intellectual abilities in offspring. The recommended optimal level of iodine intake for pregnant women is 220 µg/day [89].

Another adverse effect of iodine deficiency during pregnancy is failing to produce enough thyroid hormone, which can lead to poor neuronal proliferation and migration in the infant's brain. The results of an animal study in rats are consistent with these findings [6].

8 Role of Omega-3 on Child Neurodevelopment

Another nutrient during pregnancy and lactation that influences childhood neurodevelopment is Omega-3. However, the evidence is conflicting and does not clarify the exact amount of omega-3 fatty acids that may be beneficial to neurodevelopment [91].

The results of a review study on eight randomized controlled trials showed that omega-3 supplementation during pregnancy might have favorable effects on cognitive development in the child. However, there are insufficient studies that investigate the effects of omega-3 fatty acid supplementation during pregnancy and/or lactation on language, social-emotional, movement/physical, motor, or visual development; academic performance; or risks of attention deficit hyperactivity disorder (ADHD), *autism* spectrum disorder (ASD), anxiety, or depression [91–94].

Seven studies assessed the use of supplementation during pregnancy. Six of them did not find any association between omega-3 consumption during pregnancy and motor development in children [95–99]. One of them showed statistically significant desirable effects of supplementation on motor function in the neonate [100].

A recent Cochrane review and meta-analysis [101] found no significant effect of using omega-3 long-chain polyunsaturated fatty acids (LC-PUFA) supplementation on cognition, IQ, vision, and other neurodevelopment and growth outcomes, including language and behavior. Further studies are required to make a conclusive conclusion about the influence of giving LC-PUFA supplementation during pregnancy or lactation on neurodevelopment or visual acuity. One to two portions per week of fish/seafood high in docosahexaenoic acid (DHA) (200–300 mg/day on average) and eicosapentaenoic acid (EPA) are recommended by the American Academy of Pediatrics for lactating women. The 2020 Committee report found a favorable effect of seafood intake during pregnancy on cognitive development in young children. Consumption of at least 8 and up to 12 ounces of a variety of seafood per week (250–400 mg/day omega-3 fatty acids on average) is recommended by the Dietary Guidelines for Americans, 2020–2025. However, sources with less mercury should be selected [91].

9 Role of Docosahexaenoic Acid (DHA) on Child Neurodevelopment

DHA is the major n-3 long-chain polyunsaturated fatty acid in brain gray matter. Fifteen percent of all fatty acids in the human frontal cortex are DHA [102].

DHA plays a critical role in neurogenesis, neurotransmitter, synaptic plasticity, transmission, signal transduction in the brain, maturation of cortical astrocytes, neurovascular coupling, and glucose uptake and metabolism. It has a protective role in oxidative tissue injury,d stress in the brain, and inflammatory signaling. Maternal DHA levels in the third trimester of pregnancy are really important due to their important role in brain growth. In addition, placental development is dependent on the DHA levels; thus, the DHA status during pregnancy has become more highlighted recently. Higher DHA levels in plasma and breast milk are positively correlated with the brain's growth, development, and visual development in the neonate. The fetus is highly dependent on the maternal DHA through placental supply because DHA has limited synthesis. Taking 200 mg DHA/day is recommended for pregnant women in their third trimester to prevent the decrease of maternal DHA levels [103].

Adequate levels of DHA are required for brain development in offspring [104]. Low maternal levels of DHA in pregnant women may result in changes in placental growth and development; as a result, this can lead to behavior changes, learning problems, cognitive decline, inflammatory disease, cardiovascular disease, and inflammatory disorders [105, 106]. Thus, taking DHA supplementation can have profound effects on the mother and child's health, including the infant's neurological and visual development [103].

DHA can activate the membrane GPR40 receptor. It modulates synaptic plasticity in the brain and is responsible for driving neurogenesis. Dysfunction of GPR40 receptors can produce lipotoxicity in brain endothelial cells and neurons [107].

In addition, postnatal supplementation of DHA also improved the subsequent development of vision and brain in very low birth weight infants [108].

A large observational study on 11,875 participants showed that children of mothers who consumed a higher intake of seafood during pregnancy had higher motor skills, more significant pro-social behavior, higher verbal intelligence, and higher social development scores at 8 years of age [109].

A recent longitudinal cohort showed an association between higher DHA status during pregnancy and lactation and increasing infants' problem-solving skills at 12 months [110].

Chronic deficiency of DHA influences the process of neurodevelopment, including neuronal cell growth, differentiation, and neuronal signaling. In addition, DHA deficiency can lead to a reduction in offspring's learning, cognitive efficiency, and intellectual potential [103].

Animal studies showed that the telencephalon structure in the hippocampus region is reduced by DHA's maternal deficiency [111]. This deficiency has impacts on region-specific brain development areas, with the cerebral frontal cortex region being mostly affected, leading to hypermotor activity, reduced learning ability, and altered monoamine transmission [112].

Behavioral changes in feed intake, anxiety, and stress response in the offspring can be affected by maternal DHA deficiency [113]. The expression of the neuropeptide Y-1 receptor and glucocorticoid receptor in the prefrontal and hippocampus of the rat brain changes after chronic deficiency of n-3 fatty acids. Higher levels of DHA increase the fluidity in the structural domains of the hippocampus, prefrontal cortex, and hypothalamus regions to make the neurodevelopmental circuitry network stable, which is needed for learning and memory recognition processes [103].

In summary, adequate DHA intake and a balanced intake of n-3 and n-6 PUFAs during pregnancy and lactation are required for infant neurodevelopment.

There is a disagreement between different organizations in the world about the optimal amount of dietary intake of DHA, but the Food and Agriculture Organization (FAO) recommends a minimum dietary intake of 200 mg/day for pregnant women [114]. Some of the sources that are rich in DHA are salmon, tuna, anchovy, or mackerel [115, 116].

10 Conclusion

Human and animal studies have shown the crucial role of maternal nutrition in neurodevelopment. Providing proper and healthy nutrition during pregnancy or lactation has the potential to be a cost-effective approach in the prevention and management of neurodevelopmental disorders.

The positive effects of some micronutrients during pregnancy on a child's cognitive and behavioral outcomes are commonly acknowledged. However, future studies are needed to assess genetic heterogeneity when evaluating nutritional effects and also nutritional recommendations.

References

1. Thompson RA, Nelson CA. Developmental science and the media: early brain development. Am Psychol. 2001;56(1):5.
2. Bemanalizadeh M, Badihian N, Khoshhali M, Badihian S, Hosseini N, Purpirali M, et al. Effect of parenting intervention through "Care for child development guideline" on early child development and behaviors: a randomized controlled trial. BMC Pediatr. 2022;22(1):690.
3. Polańska K, Muszyński P, Sobala W, Dziewirska E, Merecz-Kot D, Hanke W. Maternal lifestyle during pregnancy and child psychomotor development—Polish Mother and Child Cohort study. Early Hum Dev. 2015;91(5):317–25.
4. Dórea JG. Environmental exposure to low-level lead (Pb) co-occurring with other neurotoxicants in early life and neurodevelopment of children. Environ Res. 2019;177:108641.
5. Gil A, Gil F. Fish, a Mediterranean source of n-3 PUFA: benefits do not justify limiting consumption. Br J Nutr. 2015;113(S2):S58–67.
6. Lv S, Qin R, Jiang Y, Lv H, Lu Q, Tao S, et al. Association of maternal dietary patterns during gestation and offspring neurodevelopment. Nutrients. 2022;14(4):730.
7. Tesei A, Crippa A, Ceccarelli SB, Mauri M, Molteni M, Agostoni C, et al. The potential relevance of docosahexaenoic acid and eicosapentaenoic acid to the etiopathogenesis of childhood neuropsychiatric disorders. Eur Child Adolesc Psychiatry. 2017;26:1011–30.
8. Schlögelhofer M, Amminger GP, Schaefer MR, Fusar-Poli P, Smesny S, McGorry P, et al. Polyunsaturated fatty acids in emerging psychosis: a safer alternative? Early Interv Psychiatry. 2014;8(3):199–208.
9. Robinson SM, Crozier SR, Miles EA, Gale CR, Calder PC, Cooper C, et al. Preconception maternal iodine status is positively associated with IQ but not with measures of executive function in childhood. J Nutr. 2018;148(6):959–66.
10. Bemanalizadeh M, Heidari-Beni M, Ejtahed HS, Heshmat R, Baygi F, Seif E, et al. Association of serum 25-hydroxyvitamin D concentration with anthropometric measures in children and adolescents: the CASPIAN-V study. Eat Weight Disord. 2021;26(7):2219–26. PubMed PMID: 33247367. Epub 2020/11/29. eng
11. Dhamayanti M, Noviandhari A, Supriadi S, Judistiani RT, Setiabudiawan B. Association of maternal vitamin D deficiency and infants' neurodevelopmental status: a cohort study on vitamin D and its impact during pregnancy and childhood in Indonesia. J Paediatr Child Health. 2020;56(1):16–21. PubMed PMID: 31062441. Epub 2019/05/08. eng
12. Gale CR, Robinson SM, Harvey NC, Javaid MK, Jiang B, Martyn CN, et al. Maternal vitamin D sta-

tus during pregnancy and child outcomes. Eur J Clin Nutr. 2008;62(1):68–77.
13. García-Serna AM, Morales E. Neurodevelopmental effects of prenatal vitamin D in humans: systematic review and meta-analysis. Mol Psychiatry. 2020;25(10):2468–81.
14. Freitas-Vilela AA, Pearson RM, Emmett P, Heron J, Smith AD, Emond A, et al. Maternal dietary patterns during pregnancy and intelligence quotients in the offspring at 8 years of age: findings from the ALSPAC cohort. Matern Child Nutr. 2018;14(1):e12431.
15. Bilbo SD, Tsang V. Enduring consequences of maternal obesity for brain inflammation and behavior of offspring. FASEB J. 2010;24(6):2104–15.
16. Challier J, Basu S, Bintein T, Minium J, Hotmire K, Catalano P, et al. Obesity in pregnancy stimulates macrophage accumulation and inflammation in the placenta. Placenta. 2008;29(3):274–81.
17. Heerwagen MJ, Miller MR, Barbour LA, Friedman JE. Maternal obesity and fetal metabolic programming: a fertile epigenetic soil. Am J Phys Regul Integr Comp Phys. 2010;299(3):R711–R22.
18. Hagberg H, Mallard C, Ferriero DM, Vannucci SJ, Levison SW, Vexler ZS, et al. The role of inflammation in perinatal brain injury. Nat Rev Neurol. 2015;11(4):192–208.
19. Jensen ET, van der Burg JW, O'Shea TM, Joseph RM, Allred EN, Heeren T, et al. The relationship of maternal prepregnancy body mass index and pregnancy weight gain to neurocognitive function at age 10 years among children born extremely preterm. J Pediatr. 2017;187(50–7):e3.
20. O'Shea TM, Allred EN, Kuban KC, Dammann O, Paneth N, Fichorova R, et al. Elevated concentrations of inflammation-related proteins in postnatal blood predict severe developmental delay at 2 years of age in extremely preterm infants. J Pediatr. 2012;160(3):395–401. e4
21. Caprau D, Schober M, Bass K, O'Grady S, Ke X, Block B, et al. Altered expression and chromatin structure of the hippocampal IGF1r gene is associated with impaired hippocampal function in the adult IUGR male rat. J Dev Orig Health Dis. 2012;3(2):83–91.
22. Miller NC, Georgieff MK. Maternal nutrition and child neurodevelopment: actions across generations. J Pediatr. 2017;187:10–3.
23. Malin AJ, Busgang SA, Cantoral AJ, Svensson K, Orjuela MA, Pantic I, et al. Quality of prenatal and childhood diet predicts neurodevelopmental outcomes among children in Mexico City. Nutrients. 2018;10(8):1093.
24. Ettinger AS, Lamadrid-Figueroa H, Téllez-Rojo MM, Mercado-García A, Peterson KE, Schwartz J, et al. Effect of calcium supplementation on blood lead levels in pregnancy: a randomized placebo-controlled trial. Environ Health Perspect. 2009;117(1):26–31.
25. Lohmann C. Calcium signaling and the development of specific neuronal connections. Prog Brain Res. 2009;175:443–52.
26. Neher E, Sakaba T. Multiple roles of calcium ions in the regulation of neurotransmitter release. Neuron. 2008;59(6):861–72.
27. Klein R, Sargent J, Larsen P, Waisbren S, Haddow J, Mitchell M. Relation of severity of maternal hypothyroidism to cognitive development of offspring. J Med Screen. 2001;8(1):18–20.
28. Godfrey KM, Cutfield W, Chan S-Y, Baker PN, Chong Y-S. Nutritional intervention preconception and during pregnancy to maintain healthy glucose metabolism and offspring health ("NiPPeR"): study protocol for a randomised controlled trial. Trials. 2017;18(1):1–12.
29. Chen C, Xu X, Yan Y. Estimated global overweight and obesity burden in pregnant women based on panel data model. PLoS One. 2018;13(8):e0202183.
30. Bordeleau M, Fernandez de Cossio L, Chakravarty MM, Tremblay M-È. From maternal diet to neurodevelopmental disorders: a story of neuroinflammation. Front Cell Neurosci. 2021;14:612705.
31. Gilmore JH, Knickmeyer RC, Gao W. Imaging structural and functional brain development in early childhood. Nat Rev Neurosci. 2018;19(3):123–37.
32. Naninck EF, Stijger PC, Brouwer-Brolsma EM. The importance of maternal folate status for brain development and function of offspring. Adv Nutr. 2019;10(3):502–19.
33. Caffrey A, McNulty H, Rollins M, Prasad G, Gaur P, Talcott JB, et al. Effects of maternal folic acid supplementation during the second and third trimesters of pregnancy on neurocognitive development in the child: an 11-year follow-up from a randomised controlled trial. BMC Med. 2021;19(1):1–13.
34. Bailey LB, Stover PJ, McNulty H, Fenech MF, Gregory JF III, Mills JL, et al. Biomarkers of nutrition for development—folate review. J Nutr. 2015;145(7):1636S–80S.
35. Roffman JL. Neuroprotective effects of prenatal folic acid supplementation: why timing matters. JAMA Psychiatry. 2018;75(7):747–8.
36. Caffrey A, Irwin RE, McNulty H, Strain J, Lees-Murdock DJ, McNulty BA, et al. Gene-specific DNA methylation in newborns in response to folic acid supplementation during the second and third trimesters of pregnancy: epigenetic analysis from a randomized controlled trial. Am J Clin Nutr. 2018;107(4):566–75.
37. Irwin RE, Thursby S-J, Ondičová M, Pentieva K, McNulty H, Richmond RC, et al. A randomized controlled trial of folic acid intervention in pregnancy highlights a putative methylation-regulated control element at ZFP57. Clin Epigenetics. 2019;11(1):1–16.
38. Stevens GA, Finucane MM, De-Regil LM, Paciorek CJ, Flaxman SR, Branca F, et al. Global, regional, and national trends in haemoglobin concentration and prevalence of total and severe anaemia in children and pregnant and non-pregnant women for 1995–2011: a systematic analysis of population-representative data. Lancet Glob Health. 2013;1(1):e16–25.

39. Georgieff MK. The role of iron in neurodevelopment: fetal iron deficiency and the developing hippocampus. Biochem Soc Trans. 2008;36(6):1267–71.
40. McCann S, Perapoch Amadó M, Moore SE. The role of iron in brain development: a systematic review. Nutrients. 2020;12(7):2001.
41. Santos DC, Angulo-Barroso RM, Li M, Bian Y, Sturza J, Richards B, et al. Timing, duration, and severity of iron deficiency in early development and motor outcomes at 9 months. Eur J Clin Nutr. 2018;72(3):332–41.
42. Armony-Sivan R, Zhu B, Clark KM, Richards B, Ji C, Kaciroti N, et al. Iron deficiency (ID) at both birth and 9 months predicts right frontal EEG asymmetry in infancy. Dev Psychobiol. 2016;58(4):462–70.
43. Geng F, Mai X, Zhan J, Xu L, Zhao Z, Georgieff M, et al. Impact of fetal-neonatal iron deficiency on recognition memory at 2 months of age. J Pediatr. 2015;167(6):1226–32.
44. Pallone LV, Jesus FA, Gonçalves GA, Navarra LC, Melo DG, Ferreira RA, et al. Effects of intrauterine latent iron deficiency on auditory neural maturation in full-term newborns. J Pediatr. 2020;96:202–9.
45. ElAlfy MS, Ali El-Farrash R, Mohammed Taha H, Abdel Rahman Ismail E, Ahmed MN. Auditory brainstem response in full-term neonates born to mothers with iron deficiency anemia: relation to disease severity. J Matern Fetal Neonatal Med. 2020;33(11):1881–8.
46. Tran TD, Tran T, Simpson JA, Tran HT, Nguyen TT, Hanieh S, et al. Infant motor development in rural Vietnam and intrauterine exposures to anaemia, iron deficiency and common mental disorders: a prospective community-based study. BMC Pregnancy Childbirth. 2014;14:1–11.
47. Mireku MO, Davidson LL, Boivin MJ, Zoumenou R, Massougbodji A, Cot M, et al. Prenatal iron deficiency, neonatal ferritin, and infant cognitive function. Pediatrics. 2016;138(6):e20161319.
48. Zhou SJ, Gibson RA, Crowther CA, Baghurst P, Makrides M. Effect of iron supplementation during pregnancy on the intelligence quotient and behavior of children at 4 y of age: long-term follow-up of a randomized controlled trial. Am J Clin Nutr. 2006;83(5):1112–7.
49. Doom JR, Gunnar MR, Georgieff MK, Kroupina MG, Frenn K, Fuglestad AJ, et al. Beyond stimulus deprivation: iron deficiency and cognitive deficits in postinstitutionalized children. Child Dev. 2014;85(5):1805–12.
50. Lloyd-Fox S, Blasi A, Elwell C, Charman T, Murphy D, Johnson M. Reduced neural sensitivity to social stimuli in infants at risk for autism. Proc R Soc B Biol Sci. 2013;280(1758):20123026.
51. Algarin C, Nelson CA, Peirano P, Westerlund A, Reyes S, Lozoff B. Iron-deficiency anemia in infancy and poorer cognitive inhibitory control at age 10 years. Dev Med Child Neurol. 2013;55(5):453–8.
52. Congdon EL, Westerlund A, Algarin CR, Peirano PD, Gregas M, Lozoff B, et al. Iron deficiency in infancy is associated with altered neural correlates of recognition memory at 10 years. J Pediatr. 2012;160(6):1027–33.
53. Corapci F, Calatroni A, Kaciroti N, Jimenez E, Lozoff B. Longitudinal evaluation of externalizing and internalizing behavior problems following iron deficiency in infancy. J Pediatr Psychol. 2010;35(3):296–305.
54. Lozoff B, Jimenez E, Smith JB. Double burden of iron deficiency in infancy and low socioeconomic status: a longitudinal analysis of cognitive test scores to age 19 years. Arch Pediatr Adolesc Med. 2006;160(11):1108–13.
55. Algarin C, Karunakaran KD, Reyes S, Morales C, Lozoff B, Peirano P, et al. Differences on brain connectivity in adulthood are present in subjects with iron deficiency anemia in infancy. Front Aging Neurosci. 2017;9:54.
56. Lozoff B, Smith JB, Kaciroti N, Clark KM, Guevara S, Jimenez E. Functional significance of early-life iron deficiency: outcomes at 25 years. J Pediatr. 2013;163(5):1260–6.
57. Kieliszek M. Selenium–fascinating microelement, properties and sources in food. Molecules. 2019;24(7):1298.
58. Pangrazzi L, Balasco L, Bozzi Y. Oxidative stress and immune system dysfunction in autism spectrum disorders. Int J Mol Sci. 2020;21(9):3293.
59. Lee ASE, Ji Y, Raghavan R, Wang G, Hong X, Pearson C, et al. Maternal prenatal selenium levels and child risk of neurodevelopmental disorders: a prospective birth cohort study. Autism Res. 2021;14(12):2533–43.
60. Hagmeyer S, Mangus K, Boeckers TM, Grabrucker AM. Effects of trace metal profiles characteristic for autism on synapses in cultured neurons. Neural Plast. 2015;2015:1.
61. Lyon P, Strippoli V, Fang B, Cimmino L. B vitamins and one-carbon metabolism: implications in human health and disease. Nutrients. 2020;12(9):2867.
62. Nagpal J, Mathur MR, Rawat S, Nagrath D, Lee C, Singhal A, et al. Efficacy of maternal B12 supplementation in vegetarian women for improving infant neurodevelopment: protocol for the MATCOBIND multicentre, double-blind, randomised controlled trial. BMJ Open. 2020;10(5):e034987.
63. Reischl-Hajiabadi AT, Garbade SF, Feyh P, Weiss KH, Mütze U, Kölker S, et al. Maternal vitamin B12 deficiency detected by newborn screening—evaluation of causes and characteristics. Nutrients. 2022;14(18):3767.
64. Sukumar N, Adaikalakoteswari A, Venkataraman H, Maheswaran H, Saravanan P. Vitamin B12 status in women of childbearing age in the UK and its relationship with national nutrient intake guidelines: results from two National Diet and Nutrition Surveys. BMJ Open. 2016;6(8):e011247.
65. Finkelstein JL, Layden AJ, Stover PJ. Vitamin B-12 and perinatal health. Adv Nutr. 2015;6(5):552–63.

66. Varsi K, Ueland PM, Torsvik IK, Bjørke-Monsen A-L. Maternal serum cobalamin at 18 weeks of pregnancy predicts infant cobalamin status at 6 months—a prospective, observational study. J Nutr. 2018;148(5):738–45.
67. Gramer G, Hoffmann GF. Vitamin B12 deficiency in newborns and their mothers—novel approaches to early detection, treatment and prevention of a global health issue. Curr Med Sci. 2020;40(5):801–9.
68. Mishra J, Tomar A, Puri M, Jain A, Saraswathy K. Trends of folate, vitamin B12, and homocysteine levels in different trimesters of pregnancy and pregnancy outcomes. Am J Hum Biol. 2020;32(5):e23388.
69. Wallace TC, Blusztajn JK, Caudill MA, Klatt KC, Zeisel SH. Choline: the neurocognitive essential nutrient of interest to obstetricians and gynecologists. J Diet Suppl. 2020;17(6):733–52.
70. Bekdash RA. Neuroprotective effects of choline and other methyl donors. Nutrients. 2019;11(12):2995.
71. Derbyshire E, Obeid R. Choline, neurological development and brain function: a systematic review focusing on the first 1000 days. Nutrients. 2020;12(6):1731.
72. Akison LK, Kuo J, Reid N, Boyd RN, Moritz KM. Effect of choline supplementation on neurological cognitive, and behavioral outcomes in offspring arising from alcohol exposure during development: a quantitative systematic review of clinical and preclinical studies. Alcohol Clin Exp Res. 2018;42(9):1591–611.
73. Boeke CE, Gillman MW, Hughes MD, Rifas-Shiman SL, Villamor E, Oken E. Choline intake during pregnancy and child cognition at age 7 years. Am J Epidemiol. 2013;177(12):1338–47.
74. Wu BT, Dyer RA, King DJ, Richardson KJ, Innis SM. Early second trimester maternal plasma choline and betaine are related to measures of early cognitive development in term infants. PLoS ONE. 2012;7:e43448.
75. Shaw GM, Finnell RH, Blom HJ, Carmichael SL, Vollset SE, Yang W, et al. Choline and risk of neural tube defects in a folate-fortified population. Epidemiology. 2009;714-9:714.
76. Signore C, Ueland PM, Troendle J, Mills JL. Choline concentrations in human maternal and cord blood and intelligence at 5 y of age. Am J Clin Nutr. 2008;87(4):896–902.
77. Villamor E, Rifas-Shiman SL, Gillman MW, Oken E. Maternal intake of methyl-donor nutrients and child cognition at 3 years of age. Paediatr Perinat Epidemiol. 2012;26(4):328–35.
78. Bahnfleth C, Canfield R, Nevins J, Caudill M, Strupp B. Prenatal choline supplementation improves child color-location memory task performance at 7 y of age (FS05-01-19). Curr Dev Nutr. 2019;3(Supplement_1):nzz048. FS05-01-19
79. Caudill MA, Strupp BJ, Muscalu L, Nevins JE, Canfield RL. Maternal choline supplementation during the third trimester of pregnancy improves infant information processing speed: a randomized, double-blind, controlled feeding study. FASEB J. 2018;32(4):2172.
80. Ross RG, Hunter SK, Hoffman MC, McCarthy L, Chambers BM, Law AJ, et al. Perinatal phosphatidylcholine supplementation and early childhood behavior problems: evidence for CHRNA7 moderation. Am J Psychiatry. 2016;173(5):509–16.
81. Blusztajn JK, Slack BE, Mellott TJ. Neuroprotective actions of dietary choline. Nutrients. 2017;9(8):815.
82. Strupp BJ, Powers BE, Velazquez R, Ash JA, Kelley CM, Alldred MJ, et al. Maternal choline supplementation: a potential prenatal treatment for Down syndrome and Alzheimer's disease. Curr Alzheimer Res. 2016;13(1):97–106.
83. Powers BE, Kelley CM, Velazquez R, Ash JA, Strawderman MS, Alldred MJ, et al. Maternal choline supplementation in a mouse model of down syndrome: effects on attention and nucleus basalis/substantia innominata neuron morphology in adult offspring. Neuroscience. 2017;340: 501–14.
84. Nazeri P, Tahmasebinejad Z, Pearce EN, Zarezadeh Z, Tajeddini T, Mirmiran P, et al. Does maternal iodine supplementation during the lactation have a positive impact on neurodevelopment of children? Three-year follow up of a randomized controlled trial. Eur J Nutr. 2021;60:4083–91.
85. Secretariat W, Andersson M, De Benoist B, Delange F, Zupan J. Prevention and control of iodine deficiency in pregnant and lactating women and in children less than 2-years-old: conclusions and recommendations of the technical consultation. Public Health Nutr. 2007;10(12A):1606–11.
86. Dineva M, Fishpool H, Rayman MP, Mendis J, Bath SC. Systematic review and meta-analysis of the effects of iodine supplementation on thyroid function and child neurodevelopment in mildly-to-moderately iodine-deficient pregnant women. Am J Clin Nutr. 2020;112(2):389–412.
87. Wu M, Wu D, Wu W, Li H, Cao L, Xu J, et al. Relationship between iodine concentration in maternal colostrum and neurobehavioral development of infants in Shanghai. China J Child Neurol. 2016;31(9):1108–13.
88. Levie D, Korevaar TI, Bath SC, Murcia M, Dineva M, Llop S, et al. Association of maternal iodine status with child IQ: a meta-analysis of individual participant data. J Clin Endocrinol Metabol. 2019;104(12):5957–67.
89. Zhou SJ, Condo D, Ryan P, Skeaff SA, Howell S, Anderson PJ, et al. Association between maternal iodine intake in pregnancy and childhood neurodevelopment at age 18 months. Am J Epidemiol. 2019;188(2):332–8.
90. Bayley N. Bayley scales of infant and toddler development. 2006.
91. Nevins JE, Donovan SM, Snetselaar L, Dewey KG, Novotny R, Stang J, et al. Omega-3 fatty acid dietary supplements consumed during pregnancy and lac-

tation and child neurodevelopment: a systematic review. J Nutr. 2021;151(11):3483–94.
92. Colombo J, Gustafson KM, Gajewski BJ, Shaddy DJ, Kerling EH, Thodosoff JM, et al. Prenatal DHA supplementation and infant attention. Pediatr Res. 2016;80(5):656–62.
93. Dunstan J, Simmer K, Dixon G, Prescott S. Cognitive assessment of children at age 2½ years after maternal fish oil supplementation in pregnancy: a randomised controlled trial. Arch Dis Child Fetal Neonatal Ed. 2008;93(1):F45–50.
94. Gould JF, Treyvaud K, Yelland LN, Anderson PJ, Smithers LG, McPhee AJ, et al. Seven-year follow-up of children born to women in a randomized trial of prenatal DHA supplementation. JAMA. 2017;317(11):1173–5.
95. Campoy C, Escolano-Margarit MV, Ramos R, Parrilla-Roure M, Csábi G, Beyer J, et al. Effects of prenatal fish-oil and 5-methyltetrahydrofolate supplementation on cognitive development of children at 6.5 y of age. Am J Clin Nutr. 2011;94(suppl_6):1880S–8S.
96. Keenan K, Hipwell A, McAloon R, Hoffmann A, Mohanty A, Magee K. The effect of prenatal docosahexaenoic acid supplementation on infant outcomes in African American women living in low-income environments: a randomized, controlled trial. Psychoneuroendocrinology. 2016;71:170–5.
97. Makrides M, Gibson RA, McPhee AJ, Yelland L, Quinlivan J, Ryan P, et al. Effect of DHA supplementation during pregnancy on maternal depression and neurodevelopment of young children: a randomized controlled trial. JAMA. 2010;304(15): 1675–83.
98. Mulder KA, King DJ, Innis SM. Omega-3 fatty acid deficiency in infants before birth identified using a randomized trial of maternal DHA supplementation in pregnancy. PLoS One. 2014;9(1):e83764.
99. Ramakrishnan U, Stinger A, DiGirolamo AM, Martorell R, Neufeld LM, Rivera JA, et al. Prenatal docosahexaenoic acid supplementation and offspring development at 18 months: randomized controlled trial. PLoS One. 2015;10(8):e0120065.
100. Gustafson K, Carlson S, Colombo J, Yeh H-W, Shaddy D, Li S, et al. Effects of docosahexaenoic acid supplementation during pregnancy on fetal heart rate and variability: a randomized clinical trial. Prostaglandins Leukot Essent Fat Acids. 2013;88(5):331–8.
101. Middleton P, Gomersall JC, Gould JF, Shepherd E, Olsen SF, Makrides M. Omega-3 fatty acid addition during pregnancy. Obstet Gynecol Surv. 2019;74(4):189–91.
102. Bradbury J. Docosahexaenoic acid (DHA): an ancient nutrient for the modern human brain. Nutrients. 2011;3(5):529–54.
103. Basak S, Mallick R, Duttaroy AK. Maternal docosahexaenoic acid status during pregnancy and its impact on infant neurodevelopment. Nutrients. 2020;12(12):3615.
104. McCann JC, Ames BN. Is docosahexaenoic acid, an n− 3 long-chain polyunsaturated fatty acid, required for development of normal brain function? An overview of evidence from cognitive and behavioral tests in humans and animals. Am J Clin Nutr. 2005;82(2):281–95.
105. Quinn JF, Raman R, Thomas RG, Yurko-Mauro K, Nelson EB, Van Dyck C, et al. Docosahexaenoic acid supplementation and cognitive decline in Alzheimer disease: a randomized trial. JAMA. 2010;304(17):1903–11.
106. Kabaran S, Besler HT. Do fatty acids affect fetal programming? J Health Popul Nutr. 2015;33(1):1–9.
107. Yamashima T. Dual effects of the non-esterified fatty acid receptor 'GPR40'for human health. Prog Lipid Res. 2015;58:40–50.
108. Harris W, Baack M. Beyond building better brains: bridging the docosahexaenoic acid (DHA) gap of prematurity. J Perinatol. 2015;35(1):1–7.
109. Hibbeln JR, Davis JM, Steer C, Emmett P, Rogers I, Williams C, et al. Maternal seafood consumption in pregnancy and neurodevelopmental outcomes in childhood (ALSPAC study): an observational cohort study. Lancet. 2007;369(9561):578–85.
110. Braarud HC, Markhus MW, Skotheim S, Stormark KM, Frøyland L, Graff IE, et al. Maternal DHA status during pregnancy has a positive impact on infant problem solving: a Norwegian prospective observation study. Nutrients. 2018;10(5):529.
111. Coti Bertrand P, O'Kusky JR, Innis SM. Maternal dietary (n-3) fatty acid deficiency alters neurogenesis in the embryonic rat brain. J Nutr. 2006;136(6):1570–5.
112. Carrié I, Clément M, de Javel D, Francès H, Bourre J-M. Specific phospholipid fatty acid composition of brain regions in mice: effects of n–3 polyunsaturated fatty acid deficiency and phospholipid supplementation. J Lipid Res. 2000;41(3):465–72.
113. Bondi CO, Taha AY, Tock JL, Totah NK, Cheon Y, Torres GE, et al. Adolescent behavior and dopamine availability are uniquely sensitive to dietary omega-3 fatty acid deficiency. Biol Psychiatry. 2014;75(1):38–46.
114. Joint F. Fats and fatty acids in human nutrition. Report of an expert consultation, 10–14 November 2008, Geneva 2010.
115. Emmett R, Akkersdyk S, Yeatman H, Meyer BJ. Expanding awareness of docosahexaenoic acid during pregnancy. Nutrients. 2013;5(4):1098–109.
116. Miles EA, Noakes PS, Kremmyda L-S, Vlachava M, Diaper ND, Rosenlund G, et al. The Salmon in pregnancy study: study design, subject characteristics, maternal fish and marine n–3 fatty acid intake, and marine n–3 fatty acid status in maternal and umbilical cord blood. Am J Clin Nutr. 2011;94(suppl_6):1986S–92S.

Effects of Human Milk on Neurodevelopment

Behzad Barekatain

Abstract

The fastest period of brain development occurs during the fetal and neonatal periods, and understanding the impact of factors related to brain development during the neonatal period can play a role in improving cognitive, language, motor, and social-emotional abilities. As the best source of nutrition, human milk can play a role in preventing developmental disorders and improving brain function through various mechanisms. Breast milk can positively affect the neurodevelopment of infants through three mechanisms. The first role is related to the nutritional factors in breast milk that can affect the development of the nervous system. The second role is caused by nonnutritional factors in breast milk that affect the development of the infant by regulating brain function. The third mechanism is related to how the infant is placed in the mother's arms and the stimulation of the infant's different senses. This chapter will briefly discuss these three mechanisms.

Keywords

Newborn · Breastfeeding · Neurodevelopment

1 Introduction

Infancy is one of the most important periods for brain development, and nutrition can have lasting effects on the structure and function of the developing brain.

The fastest period of brain development occurs during the fetal and neonatal periods and can determine the neurological outcomes in childhood and adulthood. Understanding the impact of factors related to brain development during infancy can play a role in improving cognitive, language, motor, and social-emotional abilities. Babies, especially premature babies admitted to neonatal intensive care units, are exposed to developmental problems due to various reasons, including prematurity, inadequate nutrition, and environmental factors. It has been proven in various studies that human milk, as the best source of nutrition, can play a role in preventing developmental disorders and improving brain function through various mechanisms. Breastfed children have better neurodevelopment, which is due to the biological and nutritional effects of breast milk, environmental effects, genetic effects, or a combination of these factors. Mothers who breastfeed their children

B. Barekatain (✉)
Department of Pediatrics, Faculty of Medicine, Isfahan University of Medical Sciences, Isfahan, Iran

Division of Neonatology, Child Growth and Development Research Center, Research Institute for Primordial Prevention of Non-Communicable Disease, Isfahan University of Medical Sciences, Isfahan, Iran
e-mail: b_barekatain@med.mui.ac.ir

© The Author(s), under exclusive license to Springer Nature Singapore Pte Ltd. 2024

R. Kelishadi (ed.), *Nature, Nurture, and Neurodevelopment*,
https://doi.org/10.1007/978-981-97-7433-3_7

usually have a higher social status. More mother–infant contact during breastfeeding provides an opportunity for more mother–infant bonding, which promotes better brain development. Hormonal effects that occur during breastfeeding affect the mother's behavior, and this affects the neural development of infants.

2 The Role of Nutritional Factors on Infant Neurodevelopment

Mother's milk, having many nutritional and non-nutritive substances, has tremendous effects on the physical and neurological development of the infant. The importance of breastfeeding includes the following: 1. It saves the lives of more than one and a half million infants every year. 2. It causes the reduction of medical expenses for the prevention and treatment of diseases and allergies. 3. It enables women to learn skills. Breastfeeding enables that it can be passed from one generation to another as a positive model. 4. It improves the emotional relationship between mother and infant. 5. It provides food security for the infant in the first 2 years of life. 6. Provides artificial feeding expenses for other necessary expenses of the family. Also, diseases such as thyroiditis, tuberculous meningitis, neonatal sepsis, acute and chronic otitis, iron deficiency anemia, constipation and abdominal pain, malnutrition, oral thrush, behavioral problems, increased blood cholesterol, gastroesophageal reflux, tooth decay and the need for Orthodontics, finger sucking, necrotizing enterocolitis, obesity, diabetes, allergies, Crohn's, leukemia, lymphoma, sudden infant death syndrome, and acute diseases (respiratory, gastrointestinal, otitis, urinary infection, sepsis, meningitis, botulism) in infants who are fed with mother's milk are less common. Also, breastfeeding has a positive effect on vaccination (BCG, Haemophilus influenzae B, polio, tetanus, and diphtheria toxoid) and positive effects on IQ and behavioral-neurological benefits. Breastfeeding has positive effects for the mother, including the prevention of maternal bleeding, weight loss after childbirth, emotional dependence and stress reduction, amenorrhea and spacing in the birth of children, prevention of cancer (breast and ovary), and a positive effect on bone density [1].

In the first few days after birth, breast milk, called colostrum, contains a high concentration of protein and antibodies. On the third to fifth days after giving birth, it is called transitional milk. Mature milk appears on the tenth day after giving birth. The main carbohydrate in breast milk is the disaccharide lactose, the amount and volume of which increase from the stage of colostrum to mature milk. Lactose in ripened milk remains constant. Lactose is widely absorbed in the intestine. A small amount of it is not absorbed, which creates a soft consistency of stool, reduces the flora of pathogenic bacteria in stool, and absorbs more minerals. Other oligosaccharides, including GOS and FOS, make up about 5% to 10% of the total carbohydrates in milk, and in addition to their nutritional role, they also play a protective role in infants [1].

Nitrogen consists of 80% protein and 20% nonprotein nitrogen compounds (NPN). In the first few weeks after delivery, the total amount of nitrogen in the milk of mothers with premature babies is higher than in the milk of mothers with full-term babies. Two to four weeks after giving birth, the amount of protein nitrogen decreases, and then its amount and composition remain constant until the time of weaning. Nonprotein nitrogen (free amino acids and urea) is more in human milk than in cow's milk (20% vs. 5%). Human milk contains 70% whey protein and 30% casein protein. In cow's milk, 18% is whey and 82% is casein. Casein is a protein with low solubility in an acidic environment, while whey protein is soluble and remains soluble even after acidification; it is easier to digest and is discharged from the stomach faster. The main protein in human milk is alpha-lactalbumin. Lactoferrin, lysozyme, and Secretory Immunoglobulin A (SIgA) are special proteins found only in human milk and act as the front line of defense against microorganisms [1].

Triglycerides make up 97–98% of fat. Cholesterol is 0.4% to 0.5%, and phospholipids are 0.6 to 0.8%. The pattern of fatty acid compounds in human milk is unique and includes long-chain polyunsaturated fatty acids (LCPUFA). Arachidonic acid (linoleic acid) and docosahexaenoic acid (linolenic acid) form the phospholipid membranes of the brain and retina and are involved in the development of vision and the nervous system. Among the components of breast milk, the total fat content has the most changes in the amount and type. Human milk fat increases throughout the day and during lactation. During one meal, lactation increases from fore milk to hind milk (two to three times). The amount of fat varies from mother to mother [1].

The concentration of vitamin K in breast milk is low, and to ensure its sufficiency, it is recommended that all infants receive 0.5 to 1 mg of vitamin K intramuscularly at birth (within the first 6 h of birth). The concentration of vitamin D in human milk is also low, and it is recommended to give 200 to 400 units of this vitamin daily to infants before the age of 2 months. Calcium and phosphorus in breast milk are constant throughout the breastfeeding period. The amount of calcium and phosphorus in mother's milk is lower compared to cow's milk, but its bioavailability and absorption are higher in human milk because it is attached to absorbable proteins and is mostly in ionized form, which is more absorbable in this case [1].

The concentration of iron, zinc, and copper in human milk decreases during lactation. It seems that the small amount of copper in breast milk is proportional to the nutritional needs of infants, but the concentration of iron and zinc in human milk may not be enough to meet the needs of infants after 6 months of age. Supplemental food and iron drops are recommended to prevent its deficiency [1].

The nutritional, cognitive, and emotional safety benefits of breast milk have been well known over the past two decades. It has been proven that breast milk has undeniable effects on visual and brain development, and the IQ of breastfed infants is higher. It is about three degrees higher than that of artificially fed infants. Numerous studies have shown the effects of long-chain unsaturated fatty acids, especially arachidonic acid and docosahexaenoic acid, on brain development. The greatest effect of these two essential fatty acids on the brain was in the third trimester of pregnancy, especially the last 2 weeks and the first 4 weeks after birth, and for this reason, the role of feeding with mother's milk is very vital for pre-term and full-term babies. A study in Sweden showed that the ratio of docosahexaenoic acid and arachidonic acid in breast milk is directly proportional to its ratio in the blood flow of the infant's brain and the growth of the head circumference at 1 and 3 months of age and the growth of the brain weight (based on the head circumference). Several other studies also confirm the relationship between feeding with mother's milk and head circumference growth, and it is interesting that most of the unsaturated fatty acids in mother's milk are influenced by the mother's diet. Another researcher and his colleagues showed in a study that even preterm infants who were fed with formulas enriched with docosahexaenoic acid and arachidonic acid were more social compared to those who were fed with colostrum [2].

Several intervening factors are effective on the IQ of infants. Some of these intervening factors include: IQ and literacy level of father and mother, socioeconomic level of the family, mother's age, home environment, mother–child relationship, mother's stress, mother's depression, infant diseases, father–child relationship, birth weight, infant delivery conditions, intrauterine growth, and intrauterine age at birth [2].

In the study conducted by Anderson and his colleagues, after controlling for confounding factors, they came to the conclusion that children who eat mother's milk have more cognitive development than those who are fed with artificial milk, and also the period of feeding with mother's milk has a higher IQ. It is direct, and the longer the breastfeeding period, the higher the IQ. This issue was more important for premature and low-birth-weight babies and showed more difference. A study by Marward and colleagues showed that children who were breastfed for more than 8 months had a higher IQ from 1.5 to 4.5 degrees. This issue in preterm babies led to a difference of about 11 degrees at the age of 8 years [2].

In order to compare the level of intelligence of children who are fed with mother's milk and children who use artificial milk, separate studies were conducted on children under 4 months and over 4 months in the city of Cleveland, USA, and all the studies showed that the longer the period of feeding with mother's milk, the more intelligent the children are. Research conducted in New Zealand showed a significant difference in behavior both at school and at home between breastfed and formula-fed children. Many other studies have been done in relation to the IQ of children who were fed with mother's milk, and all of them show an increase in IQ among breastfed children. In one of the studies conducted in England on 8-year-old children, it was shown that the children who were fed with mother's milk had a higher IQ. Another study was conducted on 9-year-old children, and the difference in IQ in this study was 12.9 degrees. Breastfed babies are more alert. This group of infants has more physical activity in the first and second weeks after birth, and their reactions are also more than other babies. A study conducted in England showed that children who are fed with mother's milk walk 2 months earlier than others, and the longer the period of feeding with mother's milk, the greater this difference [2].

Another research showed that if exclusive feeding with mother's milk continues for 6 months, the effects of developmental progress in the infant will be greater, and if the period of exclusive feeding is increased to 9 months, this progress in the child's development will still increase. However, when the exclusive feeding with mother's milk is increased to 12 months, this speed will no longer be observed. Research conducted in the state of Hawaii, America, showed that the longer the period of feeding with mother's milk, the more the child's recognition, conversation, and memory ability increase. The results of this study were unrelated to the conditions and social factors of the studied society. On the other hand, the effects of mother's milk on brain growth and brain activities in the child can also be observed in the long term, and it also has a positive effect on the strength and absorption of salts in the bones of the person in the following years. In a survey of premature babies in Taiwan, it was shown that oxygen saturation, body temperature, pulse rate, and breathing rate during breastfeeding are higher than when feeding from a bottle. Also, the possibility of apnea with a time of more than 20 s and attacks of oxygen reduction below 90% is more likely in formula-fed children than in breastfed children, which has a great impact on brain development. Long-term studies have shown that social-psychological problems and divorce are less common in those who are breastfed than in others. A study in Australia showed that female students who were fed breast milk during childhood had less anxiety and were more emotionally and psychologically balanced in puberty and adulthood. As a result, it can be said with full certainty that by promoting breastfeeding, higher intelligence and better human relations will be achieved [2].

3 The Role of Nonnutritional Factors on Infant Neurodevelopment

Nonnutritive components of breast milk include bioactive proteins (including lactoferrin, lysozyme, and SIgA), immunoglobulins (including IgM, IgD, IgE, and IgG), inflammatory and noninflammatory cytokines, free amino acids (taurine and glutamine), oligosaccharides (GOS, FOS), free fatty acids and monoglycerides, nitrogen-containing carbohydrates (bifidus factor), cellular elements (macrophages, lymphocytes, neutrophils, and epithelial cells), nucleotides (essential precursors for RNA and DNA synthesis), hormones, growth factors, and probiotics. Many of these factors play a vital role in the neurodevelopment of infants [1].

Tryptophan, found in colostrum and breast milk, is a prerequisite for the formation of serotonin, and the reduction of serotonin can be directly related to the creation of depression or violence [2].

During the period of infancy, nerve cells establish synapses with each other and create a neural network, and the occurrence of a traumatic event in this important period may have irreparable effects on movement and cognitive conditions throughout life [3].

The human body is the host of millions of living microorganisms, and the synergistic action of these microorganisms with our cells will have many effects during our lifetime and also between generations. These microorganisms, which are known as microbiota, have effective functions on the immune system, endocrine system, and nervous pathways. The regulation of brain function with gut function has been known for a long time, and the two-way nature of this relationship has also been clarified. The neuronal pathway provides communication between the gut microbiome and the brain through the afferent and efferent nerve fibers of the vagus nerve [4]. The parasympathetic nerve regulates intestinal activity, maintains systemic homeostasis, initiates anti-inflammatory activity, directly activates the central nervous system and behavior, and interacts with the HPA axis and inflammatory mediators. The bacteria present in the intestine interact with the cells of its wall and stimulate the production of peptides that activate the endings of the afferent fibers of the vagus nerve. The received signal is transmitted to the CNS, affecting behavior and activating the efferent neurons. The activation of the vagal efferent fibers, in turn, transmits inflammatory signals through the cholinergic anti-inflammatory pathway to the paramedian, which ultimately leads to the reduction of the production of inflammatory cytokines, including tumor necrosis factor-alpha. In the first year of birth, we witness a rapid progress in the expansion of the intestinal microbiome and the development of the CNS in the newborn. According to the evidence, it can be concluded that the gut microbiome plays a key role in the correct functioning of the brain, such as the regulation of brain development, neurotransmitter systems, standard signaling pathways, synaptic proteins, and emotional behaviors, including depression, mood, and smell. Therefore, a correct understanding of how the microbiome–gut–brain axis functions during infancy provides new insights into brain and neurodevelopment [5]. According to the evidence, it can be concluded that the gut microbiome plays a key role in the correct functioning of the brain, such as the regulation of brain development, neurotransmitter systems, standard signaling pathways, synaptic proteins, and emotional behaviors, including depression, mood, and smell. Therefore, a correct understanding of how the microbiome–gut–brain axis functions during infancy provides new insights into brain and neurodevelopment [6]. According to the evidence available in Jang et al.'s reviews, breast milk contains beneficial bacteria similar to the intestinal microbiome, which helps to improve the process of cognitive and physical growth, especially in premature babies [7]. With these interpretations, we can understand the importance of feeding a baby with mother's milk in the early days of birth. The studies of Deoni and colleagues showed that feeding with breast milk in infants has important consequences for cognitive and behavioral functions in children [8], as well as verbal intelligence and the improvement of oligodendrocytes for the myelination process. Defects in the myelination process can have detrimental effects on brain function. Myelin plays an important role in axon development and synapse formation, and disruption in the myelination process can cause disorders such as autism, attention deficit, hyperactivity, schizophrenia, multiple sclerosis, and congenital leukodystrophy [9].

4 The Role of Sensory Stimuli During Breastfeeding on the Infant's Neurodevelopment

Today, it has been proven that sensory stimuli, such as skin-to-skin contact, eye-to-eye contact, and a mother's whispers, which are provided while hugging an infant to feed from the breast, are necessary for the growth, development, and normal functioning of the brain [10].

In the first 8 weeks of life, skin-to-skin contact is the most important stimulus for brain development. This constant physical contact is an essential requirement for the basic structures of the brain to develop in a healthy way. After this need, the most important stimulus that the brain needs for normal development is eye contact. Both tactile and visual stimulation are best provided in the mother's arms and while feeding from the mother's breast. So, for the baby's brain, the mother is first a safe place and then becomes a safe person. The smell of the mother, both from her body and specifically from her breasts, sends specific pathways to the frontal lobe of the brain. Smell also sends signals directly to the amygdala, while direct skin-to-skin contact, particularly the deep pressure component of contact, sends messages to the same amygdala. The result is that another signal is sent by the amygdala to the prefrontal orbital cortex, and as a result of the activation of the left side, it leads to approach orientation and an anticipation of reward. Skin-to-skin contact specifically creates a left orientation. The critical period for the maturation of this pathway is 6–8 weeks. After this, eye-to-eye contact becomes important for building the next path from here [10].

Skin-to-skin contact, in particular, creates two key capacities in the infant's brain. The first is the emotional brain: the amygdala is the sorting station for all emotions, correctly deciding which situations to approach and which situations to avoid. This is emotional intelligence. The second capacity is social intelligence. The amygdala (emotional sorting station) has inputs from the face encoding area (fusiform gyrus) and the "am I safe here" decision center (gray around the gutter), and the subsequent capacity to intelligently choose between approach and avoidance. It is the core of social intelligence [10].

There is something more amazing about this birth period. The baby's behaviors send sensory signals to the mother's brain, which have a profound effect on the mother. Oxytocin—the love hormone—is produced in the mother, which increases eye-to-eye contact, and oxytocin makes even more oxytocin. In the mother's cingulate gyrus, a fear center is inhibited by high oxytocin, making her fearless in protecting her baby. By sucking, the baby stimulates prolactin, which increases milk production. Perhaps most importantly, the mother becomes more sensitive and attuned to her baby and can better understand her baby's signals, needs, and desires [10].

References

1. American Academy of Pediatrics and American College of Obstetricians and Gynecologists. Editor in chief: Richard J. Schanler, MD, FAAP, FABM. The *Breastfeeding handbook for physicians*, Third Edition.
2. Breastfeeding promotion training series, Ministry of Health, Medical Education and Medical Education of the Islamic Republic of Iran-1400.
3. Donega V, van Velthoven CT, Nijboer CH, Kavelaars A, Heijnen CJ. The endogenous regenerative capacity of the damaged newborn brain: boosting neurogenesis with mesenchymal stem cell treatment. J Cereb Blood Flow Metab. 2013;33(5):625–34.
4. Yang I, Corwin EJ, Brennan PA, Jordan S, Murphy JR, Dunlop A. The infant microbiome: implications for infant health and neurocognitive development. Nurs Res. 2016;65(1):76–88.
5. Kim H, Sitarik AR, Woodcroft K, Johnson CC, Zoratti E. Birth mode, breastfeeding, pet exposure, and antibiotic use: associations with the gut microbiome and sensitization in children. Curr Allergy Asthma Rep. 2019;19(4):22.
6. Moore RE, Townsend SD. Temporal development of the infant gut microbiome. Open Biol. 2019;9(9):190128.
7. Jang HL, Cho JY, Kim MJ, Kim EJ, Park EY, Park SA, et al. The experience of human milk banking for 8 years: Korean perspective. J Korean Med Sci. 2016;31(11):1775–83.
8. Deoni SC, Dean DC 3rd, Piryatinsky I, O'Muircheartaigh J, Waskiewicz N, Lehman K, et al. Breastfeeding and early white matter development: a cross-sectional study. NeuroImage. 2013;82:77–86.
9. Spader HS, Ellermeier A, O'Muircheartaigh J, Dean DC 3rd, Dirks H, Boxerman JL, et al. Advances in myelin imaging with potential clinical application to pediatric imaging. Neurosurg Focus. 2013;34(4):E9.
10. Wang Y, Zhao T, Zhang Y, et al. Positive effects of kangaroo mother care on long-term breastfeeding rates, growth, and neurodevelopment in preterm infants. Breastfeed Med. 2021;16(4):282–91.

Sleep and Neurodevelopment

Zahra Parsapour, Maryam Bemanalizadeh,
Vida Imani, Marc N. Potenza, and Amir H. Pakpour

Abstract

Sleep is an essential aspect of early childhood development. Adequate sleep is critical for various physiological and cognitive processes, such as memory consolidation, emotional regulation, and attention regulation. Infants' and children's sleep needs differ according to their age. Inadequate and low-quality sleep in infants may not only endanger their mental health but also lead to long-term adverse effects on cognitive, language, and even sensory-motor abilities. Therefore, knowing potential risk factors for sleep disorders is a foundation for preventing and then treating these problems in infants and assisting their parents in these efforts. In this chapter, we have provided an overview of normal sleep in infants and toddlers and its effects on neurodevelopment. We next evaluated the state of sleep disorders among infants and children and described interventions.

Keywords

Rapid eye movement · Sleep deprivation · Neurodevelopmental delays · Developmental disorders · Sleep

1 Normal Sleep in Infants and Children

Sleep is considered a basic human need that plays a key role in children's daily lives. Sleep, as a complex physiological and behavioral process, is related to many functions, including but not limited to improving attention, learning and memory, promoting normal brain development, and strengthening immune function. Sleep not only supports physical health but also leads to maintaining human cognitive performance [1]. Studies also suggest important relationships between sleep and mental health. For example, sleep-related concerns have been reported in children with symptoms of depression and anxiety more so than in other nonaffected peers [2]. Therefore, healthy sleep appears to relate importantly to

Z. Parsapour · M. Bemanalizadeh
Child Growth and Development Research Center, Research Institute for Primordial Prevention of Non-Communicable Disease, Isfahan University of Medical Sciences, Isfahan, Iran

V. Imani (✉)
Pediatric Health Research Center, Tabriz University of Medical Sciences, Tabriz, Iran

Sleep Disorders Unit, Department of Neurology, Acibadem University, Istanbul, Turkey

M. N. Potenza
Department of Psychiatry, Yale University School of Medicine, New Haven, USA

A. H. Pakpour
Department of Nursing, School of Health and Welfare, Jönköping University, Jönköping, Sweden

© The Author(s), under exclusive license to Springer Nature Singapore Pte Ltd. 2024

R. Kelishadi (ed.), *Nature, Nurture, and Neurodevelopment*,
https://doi.org/10.1007/978-981-97-7433-3_8

general human health by improving both physical and mental health.

The early years of human life are considered critical due to rapid and extensive changes in development. Thus, sleep deprivation for fetuses and newborns may have destructive and long-term effects on sensory development [3]. In this section, we briefly summarize aspects of normal sleep and health in children.

1.1 Normal Sleep in the Early Years of Life

As mentioned above, sleep is a behavioral state that is related to human physical and mental health. Categories of sleep and wakefulness include three states: wakefulness, nonrapid eye movement (NREM) sleep, and rapid eye movement (REM) sleep.

NREM sleep is considered a restorative phase of sleep during which brain activity is low. However, during this stage, the brain is able to perform some important functions and maintain body movements.

On the other hand, REM sleep is a stage of sleep where the brain is more active and produces waves of low voltage and high frequency. REM sleep not only helps to consolidate memory but may also play a significant role in the development of the central nervous system. This stage is mainly known for dreaming.

In general, sleep patterns and stages differ at specific ages, and understanding these changes is important to help ensure adequate sleep quality and overall health. Infants spend more than half of their long hours of sleep [16–18] in the REM stage, while premature infants spend more of their sleep (80%) in this stage. Infants need more sleep than adults, and their sleep cycle includes more REM than NREM sleep. NREM sleep is subdivided into three stages, including stages 1, 2, and 3. Stage 1 (N1), which is also called light sleep, is the first stage of NREM sleep when the infant wakes up and falls asleep. The characteristics of this stage include drowsiness, muscle twitches, and short periods of wakefulness, and the brain waves in this stage are slower and more irregular than when awake. In the second stage (N2), muscle activity and awareness of the external environment decrease. Brain waves become slower and more synchronized during this stage, and sometimes rapid brain activity is called sleep spindles and K-complexes.

The final stage is classified as deep sleep or slow-wave sleep (N3), where the brain waves become very slow, known as delta waves. It is at this stage that children have difficulty waking up and may experience sleepwalking, bedwetting, or night terrors. This stage is important for physical repair and growth, as well as for memory consolidation. The stage of deep sleep in infants is shorter, and they spend more time in the lighter stages of sleep.

Newborn infants have states of active sleep, quiet sleep, and indeterminate sleep. Over time (especially 6 months after birth), quiet sleep turns into NREM sleep and active sleep into REM sleep.

Newborn children do not have a circadian rhythm until 3 months old, so sleep can easily happen during their day or night. The duration of sleep needed by a full-term baby is between 16 and 18 hours, of which 2.5 to 4 hours are continuous sleep. In the beginning, even the waking pattern is irregular, which typically gradually adjusts with breastfeeding time [4].

By the time the infant is 1 year old (from the third month of birth), not only does the circadian rhythm appear, but the circadian and ultraradial processes become more like adults. Social environmental cues, such as breastfeeding time and night routines, can help speed up this process. From the age of 6 months, the infant sleeps continuously for about 6 hours, and until the age of 1 year, he/she sleeps 14 to 15 hours a day in total [4]. As long as the child grows up, his/her need for sleep decreases, so that until the age of 2 years, he/she sleeps about 13 hours a day. This amount at the ages of 4 and 5 years decreases to about 12 and 11 hours, respectively. Along with reductions in the total duration of sleep, the number of daily naps also decreases from two naps to one nap until about the age of 18 months. Also, REM sleep decreases by about 50% from infancy to adulthood, so that after 6 months after birth, sleep-wake patterns include REM and NREM similar to those of adults. There have not been many studies on the relationship between REM and NREM, but most have found an important relationship with cognitive functions and even growth outcomes [5, 6].

2 Sleep and Development

2.1 Sleep and Memory

Sleep may play an important and vital role in the formation of memory and learning. As infants' brains are developing, more sleep is needed. Various studies have shown that a good night's sleep for infants can be associated with memory consolidation and better working memory. These relationships have not only been investigated in night sleep, but their relationship with the infant's daily naps has also been observed [7]. For example, in a study conducted on infants aged 6 months to 12 months, it was found that taking a daily nap for 30 minutes after learning can greatly help improve their memory performance [8].

During sleep, the process of memory consolidation, which leads to the formation of long-term memory, intensifies. During infancy and childhood, which is considered an important period of brain development, new brain synapses are generated with instances of learning. However, this process may not continue beyond the age of 2–3 years, because the capacity of the brain accommodates a limited number of synapses, and "synapse overload" needs to be avoided. Therefore, synaptic pruning can help remove weak and unnecessary synapses. Sleep deprivation can lead to disruptions in synaptic pruning and, eventually, disruption of processes promoting memory consolidation. There exist relatively few studies in this area, and these studies have largely been conducted preclinically [9].

2.2 Sleep and Psychosocial Health

Sleep may contribute importantly to emotional regulation and mental health. Currently, important roles of sleep deprivation and sleep disorders in the development of mental disorders such as anxiety and depression are documented [10]. Sleep deprivation can not only lead to a decrease in cognitive functions but also increase the experience of negative emotions [11]. Daily stressful events can not only lead to an increase in REM-density and REM-related sleep duration but also decrease the quality of sleep [12]. In a review, it was found that moods can be affected much more than cognitive and motor functions in people who are deprived of sleep. In a study of 84 infants aged 30 to 36 months, sleep deprivation was related to more negative emotional responses and less self-regulation. Changes in REM sleep have been associated with changes in mood and emotions, and daily negative emotions have been effective in reducing the duration and increasing the density of REM sleep [13]. In addition, sleep deprivation may reduce a person's motivation to seek social activities to find rewards. These factors may significantly impact people's social interactions [14].

In terms of social functioning, adequate and high-quality sleep can contribute to optimal performance. Sleep duration has an important relationship with peer relationships, cognitive skills, and even peer acceptance [15]. Insufficient sleep can increase behavioral problems and lead to irritability, short attention spans, and difficulties in regulating impulses and emotions in preschool children. For example, Pressman and Imber found that parents of 2- to 13-year-old children who had irregular bedtimes were more likely to report behavioral problems such as tantrums or meltdowns [16]. In another study, Sadeh et al. examined the sleep of 43 infants aged 1 year over a 3-year period. They found that poor quality sleep may compromise attention regulation and increase behavioral problems [17]. Apart from a series of psychological and behavioral problems that a lack of sleep may cause for infants, their families are also prone to mental health problems. Mothers of infants who have sleep problems may use physical punishment more frequently and experience disturbing thoughts of harming them [18, 19].

2.3 Sleep and Language Development

Sleep can play an especially effective role in language development during infancy. For example, Dionne and colleagues conducted a study to examine the relationship between sleep consolidation (an index used to assess sleep fragmentation) in infants at 6, 18, and 30 months of age and language skills. They found that, compared to other infants, infants with language delays at 60 months had less sleep consolidation at 6 and 18 months. Therefore, poor sleep may be an important risk factor for language delays [20]. This issue has been investigated and confirmed in other studies [21]. Shorter sleep, waking up more at night, and poorer sleep quality may slow down language development in infants. However, the role of environmental and genetic factors in this relationship warrants consideration. Socioeconomic status (SES) may play an important role in language development in infants. A recent longitudinal study not only emphasizes the role of nighttime sleep in the ability to acquire vocabulary but also acknowledges that low SES may make this role less prominent [22]. Despite the fact that several studies have been conducted after the age of 6 months regarding the development of language and sleep in infants, none have investigated this relationship before the age of 6 months.

2.4 Sleep and Growth

During the first 6 months of an infant's life, there are many changes in development and growth that may affect sleep. During this period, infants typically have long periods of REM sleep and short sleep cycles, the circadian rhythm begins to develop, and very rapid physical changes occur. The duration of sleep may be related to anthropometric indicators. For example, sleeping less than 12 hours in infants increases the likelihood of obesity and overweight at the age of 3 years. On the other hand, shorter durations (less than 12 hours) of sleep for 3-month-old infants are related to shorter body length. However, such findings are often based on parents' self-report criteria, which may be subject to recall bias. Also, the role of other risk factors such as parental weight, gender, and inactive lifestyles should be investigated [23]. Nonetheless, length and height may be related to infants' sleep. Longer sleep duration and more frequent naps may promote the length and height of infants. Short growth in underprivileged societies like Nepal, along with iron deficiency in infants, has been associated with sleep patterns such as reduced sleep duration, sleepwalking, and reduced napping [24]. Sleep patterns may affect not only infant growth but also fetal growth, as fetuses with intrauterine growth restriction have less active sleep and more quiet and indeterminate sleep [25].

2.5 Sleep and Sensorimotor Development

While there have not been many studies on motor development and sleep in infants, some suggest that sleep may play an important role in consolidating motor skills and the experiences gained during the infant's waking hours. In a study by Copes and Rosentswieg, the effect of sleep deprivation on students' motor performance was investigated. The results indicated that sleep deprivation may lead to lower motor performance [26]. Also, Berger and Scher investigated the potential effects of napping on motoric problem-solving in infants. The obtained results showed that infants who napped more frequently performed better in solving motor problems compared to infants who did not nap [27]. However, more studies are needed to examine the longitudinal roles of sleep in the motor development of infants.

3 Sleep in Children with Neurodevelopmental Disorders

3.1 Etiologies and Risk Factors

Sleep disturbance (sleep problems and sleep disorders) is not a diagnosis or precise condition, similar to "pain," for which there may be multiple underlying etiologies. Hence, it is important to detect the underlying factors to target the main sources of sleep disturbances and to provide appropriate interventions [28].

There are several underlying etiologies for sleep disturbances in children with neurodevelopmental disorders (NDDs) or developmental delays (DDs). A complex variety of biological, physiopathological, anatomical, psychological, and environmental factors may contribute to sleep disturbances among individuals with NDDs or DDs (Fig. 1) [29].

Anatomical malformations or injuries to parts of the brain (e.g., thalamus, basal forebrain, brain stem), as well as the maldevelopment of hormonal and neurotransmitter systems at neuronal junctions influencing wakefulness or sleep, may contribute to developing sleep disturbances in individuals with NDDs or DDs [30]. For instance, melatonin, a sleep cycle regulator that is synthesized in the pineal body, has been demonstrated to be decreased in a wide range of NDDs, including Smith-Magenis syndrome, Angelman syndrome, and autism spectrum disorders. Individuals with autism spectrum disorders may also have low urine, serum, or plasma levels of 6-sulfatoxymelatonin (a melatonin metabolite) [31–33]. Hyper-arousal states [34], hypersensitivity to specific sensory stimuli (i.e., tactile, auditory, or visual) [35], and increased levels of cognitive activity (thoughts and worries) [36] in individuals with autism spectrum disorders may result in delayed sleep onset.

Clock genes, such as human PER1(hPER1), hPER2, and hPER3, and their interactions may influence both sleep and brain development [37].

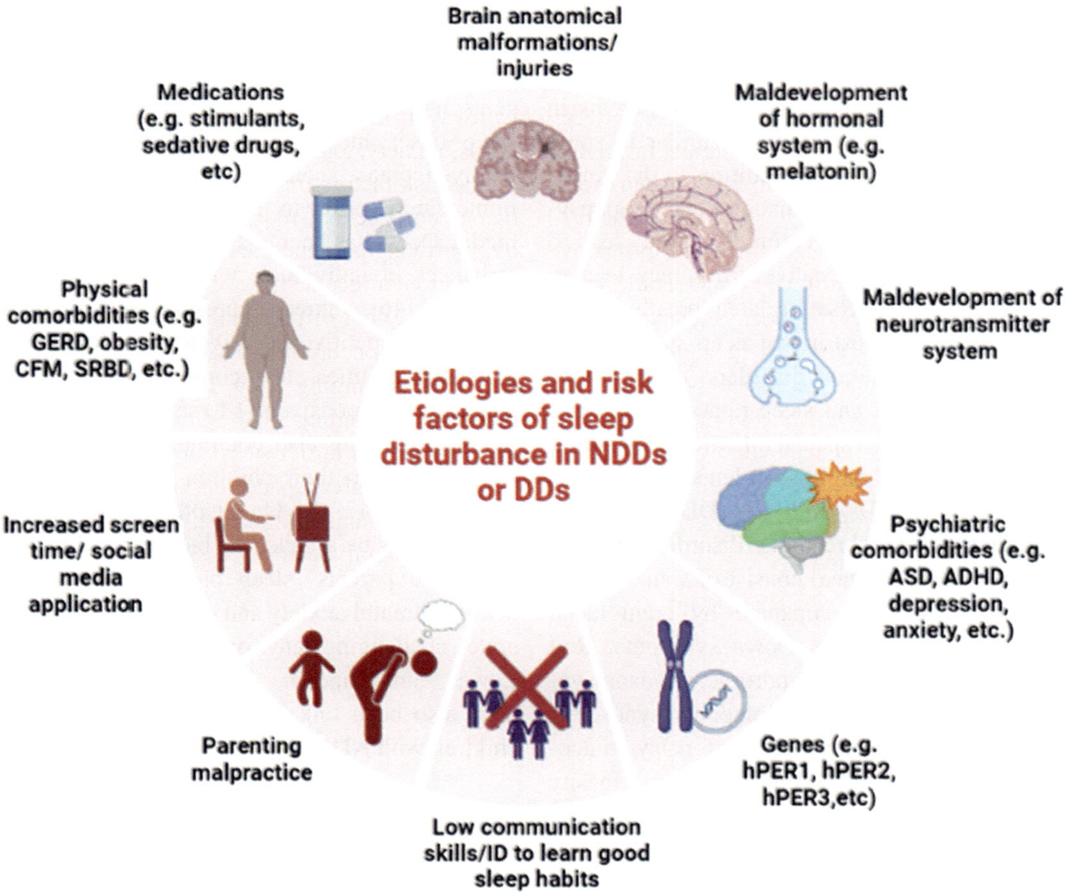

Fig. 1 Potential etiologies and risk factors of sleep disturbances in neurodevelopmental disorders (*NDDs* Neurodevelopmental disorders, *DDs* Developmental delays, *ASD* Autism spectrum disorder, *ADHD* Attention deficit hyperactivity disorder, *hPER* human PER, *ID* Intellectual disability, *GERD* Gastroesophageal reflux disorder, *CFM* Craniofacial malformation, *SRBD* Sleep-related breathing disorder)

In addition, some genes, such as the acetylserotonin O-methyltransferase gene and the cytochrome P450 1A2 gene, may influence not only synaptic formation and hemostasis but also the control of sleep-wake cycles [38, 39].

Specific co-occurring physical and psychiatric disorders or conditions that may disturb normal sleep cycles and circadian rhythms in children with NDDs or DDs include, but are not limited to, epilepsy, gastroesophageal reflux, asthma, allergies, craniofacial abnormalities, sleep-related breathing disorders, obesity, pain, arousal and sensory dysregulation, depression, bipolar disorder, psychosis, anxiety, and attention deficit hyperactivity disorder [40, 41]. Seizures (particularly of the frontal and temporal lobe) may interfere with sleep cycles. They may directly affect children with nighttime convulsive seizures in Landau-Kleffner and Lennox-Gaustaut syndromes. Furthermore, some antiepileptic medications may complicate sleep problems in children [42]. This scenario is similar to some psychiatric conditions. In addition to the potential effects of psychiatric disorders on sleep, psychotropic drugs, such as stimulants, may lead to wakefulness, while sedative drugs may lead to oversleeping [43]. Sleep-related breathing disorders, including obstructive sleep apnea, sleep-related hypoventilation disorders, central sleep apnea syndromes, and sleep-related hypoxemia disorder, often need special attention due to their wide variety and high prevalence in children with NDDs or DDs. NDDs or DDs associated with sleep-related breathing disorders (mainly obstructive sleep apnea) consist of some genetic diseases mostly accompanied by craniofacial malformations, such as Down syndrome, Rett syndrome, Prader-Willi syndrome, Crouzon syndrome, Apert syndrome, Goldenhar syndrome, Pierre Robin sequence, cerebral palsy, mucopolysaccharidoses, and achondroplasia. Obesity and chronic nasal obstruction may promote sleep disturbances in individuals with NDDs or DDs [44]. In general, the type of sleep disturbance presents similarly as in the general population. However, sleep problems associated with NDDs or DDs can be more prevalent, severe, and persistent, especially if not identified early and treated appropriately. Since many co-occurring disorders relate importantly to sleep disturbances in NDDs or DDs, assessment and treatment typically benefit from comprehensive clinical perspectives.

Along with clinicians' roles in the early identification of sleep problems in children with NDDs or DDs, parents may also contribute importantly, not only in detection but also in preventing the development of sleep problems. Some poor parenting practices, such as a lack of regular routines for helping prepare their children for bed, poor boundaries and limits when their children resist and refuse to go to sleep, and unintentional reinforcement of their children's behavior by paying inappropriate attention, may make their children's sleep problems worse. Hence, parenting interventions may help manage bedtime and sleep problems in children [30].

Learned behaviors may be important causes of prolonged night wakings and delayed sleep onset in children. For instance, in children and adolescents with autism spectrum disorders, delayed sleep onset and difficulty falling asleep may involve increased screen time during the day and immediately prior to bedtime [45]. Bedroom media access has been associated with sleep disturbances in individuals with autism spectrum disorders [46]. Anxiety, depression, self-injurious behaviors, repetitive stereotypic behaviors, intellectual disabilities, low communication skills, and limited responsiveness to social and environmental cues may also complicate parents' tendencies to help their children progress toward proper sleep [47, 48]. On the other hand, sleep in children may be affected by bidirectional factors related to parents' sleep patterns and mental health. Parental anxiety and depression may generate challenging environments for children. Lower family income and parental education have also been linked to sleep disturbances in children with NDDs or DDs [49].

3.2 Specific Neurodevelopmental Disorders

3.2.1 Autism

Autism spectrum disorders include multiple NDDs and start from the neonatal stages. Individuals with autism experience varying degrees of dissociation from the environment and relatives, and they have restricted or repetitive behaviors or interests. Sleep disorders in individuals with autism may not only affect daily functioning and quality of life but also lead to poor family functioning and increased stress.

Delayed onset of falling asleep, interrupted sleep, and low-quality sleep are relatively common in individuals with NDDs. Parasomnia is prevalent, but often, because of speech disorders, it may go underdiagnosed. Epilepsy is also relatively common, and in cases of structural sleep disorders, it may be even more frequent and lead to more sleep problems [50, 51].

3.2.2 Down Syndrome

This syndrome results from a total or partial extra chromosome 21. Down syndrome features include cognitive, behavioral, and physical symptoms such as a flat nose bridge, short neck, slanted eyes that point upward, palmar creases, and shorter-than-average height.

In people with Down syndrome, sleep disorders such as obstructive sleep apnea are more prevalent and less frequently diagnosed because of anatomical upper airway problems like adenoid hypertrophy, macroglossia, and midfacial and mandibular hyperplasia. In addition, muscular hypotonia and obesity are common and may contribute to poor sleep quality, decreased REM periods, and the maintenance of insomnia. If obstructive sleep apnea is not diagnosed and treated, an increased risk may exist for cardiovascular diseases such as left ventricular hypertrophy, pulmonary hypertension, and obstructive sleep apnea

In people with Down syndrome, more fragmented sleep, less total sleep time, and more stage 1 sleep may aggravate mood disorders like depression, irritability, and hyperactivity. Additionally, negative effects on concentration and learning may occur. These factors represent reasons for early interventions targeting sleep disorders [48–54].

3.2.3 Angelman Syndrome

Angelman syndrome involves developmental delays, glossoceles, microcephaly, spaced teeth, and an ataxic gait. Individuals may experience sleep problems such as delayed onset of sleep, interrupted sleep, laughter and screaming during sleep, sleepwalking, bruxism, day-and-night rhythm disorders, and seizures while sleeping. Periodic limb movement disorder is also a potential concern [50–54].

3.2.4 Prader-Willi Syndrome

Initial symptoms in individuals with Prader-Willi syndrome include eating disorders and hypotonia. Prader-Willi syndrome may progress to include developmental disorders and hyperplasia. Infant polysomnography has revealed reductions in REM latency and sleep-onset REM periods, which may explain narcolepsy-like symptoms [52].

3.3 Diagnosis

3.3.1 Sleep Measures

Several tools are used to assess sleep disorders in children, each with advantages and limitations. Several main methods include the following.

3.3.2 Polysomnography

Like other children, the gold standard method for measuring some sleep problems, including obstructive sleep apnea, periodic limb movement, and parasomnias, is polysomnography.

Polysomnography uses an electroencephalogram (EEG), an electrooculogram (EOG), and a chin electromyogram (EMG) to measure brain activity and body movements during sleep. Chest and abdominal belts and nasal thermistors measure respiration and respiratory effort. However, in children with neurodevelopmental problems, it is sometimes difficult to perform polysomnography due to stress related to being in a new environment, sleeping away from home, and high sensory sensitivity to the electrodes used. In addition, polysomnography may be costly and time-consuming [55, 56].

3.3.3 Actigraphy

Actigraphy is another method for assessing sleep disorders in children. Actigraphs are small, computerized devices that may be worn on one's nondominant wrist or ankle. They provide a general estimate of sleep-wake cycles by measuring activity levels. In order to get a representative sample of data, children should wear actigraphs for at least 1–2 weeks. Actigraphy is less invasive and less expensive than polysomnography and is mostly used in research studies. In children with NDDs, actigraphy has generated reliable results compared to polysomnography [57]. In addition, actigraphy is often more acceptable to children with NDDs because it can be performed at home [57]. Among the problems associated with actigraphy, its ability to distinguish between sleep and quietly lying may lead to decreased accuracy in measuring total sleep time, which can be mitigated by adding videosomnography [58].

3.3.4 Caregiver/Self-Report

Considering problems related to objective assessments of children with NDDs, high costs, and limited access, more subjective methods such as sleep questionnaires and diaries are often used in screening for sleep problems. Widely used sleep questionnaires in screening for sleep problems in children with NDDs include the BEARS questionnaire (B=Bedtime Issues, E = Excessive Daytime Sleepiness, A = Night Awakenings, R = Regularity and Duration of Sleep, S=Snoring) and the Children's Sleep Habits Questionnaire (CSHQ) [59, 60]. The BEARS questionnaire is a validated tool used to screen for common sleep problems in children. The BEARS has been found to be a reliable and valid measure of sleep problems in children aged 2–18 years. The CSHQ has been widely used for assessing a wide range of sleep behaviors and problems, including bedtime resistance, sleep onset delay, night wakings, sleep duration, and sleep-disordered breathing in children (aged 4–10 years). The validity and reliability of the CSHQ have been well established. Two questionnaires, the Behavioral Evaluation of Disorders of Sleep Scale (BEDS) and the adapted Simonds Parraga Sleep Questionnaire, were specifically designed for children with NDDs [61, 62]. The BEDS has 26 items that cover a wide range of sleep behaviors and issues, including sleep onset delay, night wakings, sleep-related anxiety, and sleep-disordered breathing. The BEDS has been validated in several studies on children with NDDs aged 2–18 years. The Simonds Parraga Sleep Questionnaire was originally developed to measure sleep problems in children. However, it has been adapted for use with children with NDDs.

3.4 Treatment

In the treatment of sleep disorders in children with NDDs, co-occurring sleep conditions, such as sleep breathing disorders, restless legs syndrome, periodic limb movement disorders, and rhythmic movement disorder, should first be diagnosed. Next, the most appropriate treatment should be considered, which generally includes an extensive discussion. In this part, the main focus should be on the treatment of insomnia.

Behavioral and pharmacologic therapies are the main methods for treating these conditions. The first line of treatment involves behavioral therapy. If there is no improvement with behavioral therapy, pharmacotherapy may be used.

3.4.1 Behavioral Interventions

In the diagnosis and treatment of insomnia, especially in children with NDDs, a detailed history should first be taken. Environmental factors affecting sleep, including noise and ambient light, bedroom conditions, children's temperaments, and bedtime routines, should be evaluated. The purpose of behavioral therapy is to improve self-soothing skills in children, which relate to children's abilities/tendencies to fall asleep and return to sleep independently [63].

To improve sleep, sleep hygiene should be evaluated and improved as indicated. This may include controlling environmental factors such as reducing light and noise near bedtime, keeping the bedroom relatively cool, limiting visual and auditory stimuli (such as electronic devices) before sleep, having a successful and consistently followed bedtime routine, and maintaining a consistent sleep/wake schedule [64].

Like other children, in those with NDD, various behavioral strategies may help with sleep. Some of these strategies include the following.

Positive routines and bedtime fading: In this method, parents develop enjoyable and calming associations before bedtime in wind-down periods, with rewards for effort. Once sleep is established, bedtimes are moved earlier by 15 minutes over successive nights until the bedtime goal is reached. Bedtimes are delayed by 30 minutes if it takes more than 30 minutes to fall asleep [65].

Graduated extinction: Several methods warrant mention. One method involves parents being instructed to ignore bedtime crying for specified periods [66]. Another method includes limit setting (setting firm and consistent limits), camping out (sleeping in the child's room and leaving it gradually), and checking out (leaving the child's room for short periods and re-entering) [66].

Cognitive behavioral therapy (CBT) may be used in older children. CBT is a first-line psychological treatment that focuses on identifying and changing negative patterns of thought and behavior that contribute to psychological distress. According to CBT, thoughts, feelings, and behaviors are interconnected. Negative or distorted thoughts may lead to negative emotions and behaviors, while positive and realistic thoughts may lead to positive emotions and behaviors. CBT for Insomnia (CBT-I) is a specific form of CBT that has been developed for treating sleep problems in children and adults. CBT-I has been found to be an effective therapeutic approach in treating insomnia among children and adolescents [67].

Regular and continuous implementation of these strategies by parents or healthcare providers often improves sleep onset latency and total sleep time and reduces awakenings during the night. These changes may emerge slowly and improve over weeks to months if the strategies are conducted regularly.

3.4.2 Pharmacotherapeutic Management

Behavioral techniques may be difficult to implement and may take weeks or months to show benefits. Thus, medications may be considered for children not demonstrating significant improvement from behavioral interventions. No medications with indications for sleep are approved by the FDA, given limited controlled studies. The European Medicines Agency has approved only one medication (pediatric prolonged-release melatonin) for pediatric insomnia [68].

Melatonin is a neurohormone that has been implicated in controlling circadian sleep-wake rhythms. This medication is prescribed by many pediatricians for sleep-onset insomnia, delayed sleep phase, and nighttime awakenings [69]. Studies have investigated melatonin in children with autism spectrum disorders, Angelman syndrome, or those with NDDs, with a dose of 1–15 mg taken 30–60 minutes before sleep, with or without behavioral therapy. Findings suggest that melatonin is effective in treating insomnia in children. However, in multiple studies, melatonin did not have a significant effect on decreasing sleep latency, and there was nonsignificant improvement of total nighttime sleep [70].

Other medications used to treat insomnia in children with NDDs may include ramelteon [71], antihistamines [72], clonidine [73], guanfacine [74], benzodiazepines [75], and gabapentin [76]. Additionally, some studies have suggested possible roles for antidepressants, chloral hydrate, orexin antagonists, atypical antipsychotics, and non-benzodiazepine sedative-hypnotics [77].

4 Conclusion

In conclusion, sleep is a vital aspect of early childhood development. Neglecting its important role may lead to disruptions in developing physiological and cognitive processes. Normal sleep in the early years of life is characterized by regular sleep-wake cycles, deep sleep stages, and shorter periods of REM sleep. Sleep disturbances may affect the overall health of infants and children. Infants and children with NDDs appear particularly vulnerable to sleep disturbances, which may exacerbate other symptoms of the disorders and disrupt family routines. Diagnosis and management of sleep disorders in infants and children with NDDs require careful attention to the unique needs and conditions of the child. Behavioral interventions and medication management are often effective treatment options that may help develop healthy sleep in children. Understanding the relationship between sleep and neurodevelopment may help healthcare providers and parents develop effective strategies to promote healthy sleep in infants and children.

References

1. Colrain IM. Sleep and the brain. Neuropsychol Rev. 2011;21:1–4. https://doi.org/10.1007/s11065-011-9156-z.
2. Gregory AM, Sadeh A. Annual research review: sleep problems in childhood psychiatric disorders—a review of the latest science. J Child Psychol Psychiatry. 2016;57(3):296–317.
3. Graven SN, Browne JV. Sleep and brain development: the critical role of sleep in fetal and early neonatal brain development. Newborn Infant Nurs Rev. 2008;8(4):173–9.
4. Davis KF, Parker KP, Montgomery GL. Sleep in infants and young children: part one: normal sleep. J Pediatr Health Care. 2004;18(2):65–71.
5. Arditi-Babchuk H, Feldman R, Eidelman AI. Rapid eye movement (REM) in premature neonates and developmental outcome at 6 months. Infant Behav Dev. 2009;32(1):27–32.
6. Weisman O, Magori-Cohen R, Louzoun Y, Eidelman AI, Feldman R. Sleep-wake transitions in premature neonates predict early development. Pediatrics. 2011;128(4):706–14.
7. Hupbach A, Gomez RL, Bootzin RR, Nadel L. Nap-dependent learning in infants. Dev Sci. 2009 Nov;12(6):1007–12.
8. Seehagen S, Konrad C, Herbert JS, Schneider S. Timely sleep facilitates declarative memory consolidation in infants. Proc Natl Acad Sci. 2015;112(5):1625–9.
9. Zhou Y, Lai CSW, Bai Y, Li W, Zhao R, Yang G, Frank MG, Gan W-B. REM sleep promotes experience-dependent dendritic spine elimination in the mouse cortex. Nat Commun. 2020;11:4819.
10. Alfano CA, Gamble AL. The role of sleep in childhood psychiatric disorders. Child Youth Care Forum. 2009;38:327–40.
11. Kahn M, Sheppes G, Sadeh A. Sleep and emotions: bidirectional links and underlying mechanisms. Int J Psychophysiol. 2013;89(2):218–28.
12. Vandekerckhove M, Cluydts R. The emotional brain and sleep: an intimate relationship. Sleep Med Rev. 2010;14(4):219–26.
13. Berger RH, Miller AL, Seifer R, Cares SR, LeBourgeois MK. Acute sleep restriction effects on emotion responses in 30- to 36-monthold children. J Sleep Res. 2012 Jun;21(3):235–46.
14. Steptoe A, O'Donnell K, Marmot M, Wardle J. Positive affect, psychological well-being, and good sleep. J Psychosom Res. 2008;64(4):409–15.
15. Goodlin-Jones B, Tang K, Liu J, Anders TF. Sleep problems, sleepiness and daytime behavior in preschool-age children. J Child Psychol Psychiatry. 2009;50(12):1532–40.
16. Pressman RM, Imber SC. Relationship of children's daytime behavior problems with bedtime routines/practices: a family context and the consideration of faux-ADHD. Am J Fam Ther. 2011;39(5):404–18.
17. Sadeh A, De Marcas G, Guri Y, Berger A, Tikotzky L, Bar-Haim Y. Infant sleep predicts attention regulation and behavior problems at 3–4 years of age. Dev Neuropsychol. 2015;40(3):122–37.
18. Johnson CM. Infant and toddler sleep: a telephone survey of parents in one community. J Dev Behav Pediatr. 1991;12(2):108–14.
19. Fairbrother N, Woody SR. New mothers' thoughts of harm related to the newborn. Arch Women Ment Health. 2008;11(3):221–9.
20. Dionne G, Touchette E, Forget-Dubois N, Petit D, Tremblay RE, Montplaisir JY, Boivin M. Associations between sleep-wake consolidation and language development in early childhood: a longitudinal twin study. Sleep. 2011;34(8):987–95.
21. Botting N, Baraka N. Sleep behaviour relates to language skills in children with and without communication disorders. Int J Develop Disabilities. 2018;64(4–5):238–43.
22. Knowland VCP, Berens S, Gaskell MG, Walker SA, Henderson L. Does the maturation of early sleep patterns predict language ability at school entry? J Child Lang. 2022;49(1):1–23.
23. Spruyt K. A review of developmental consequences of poor sleep in childhood. Sleep Med. 2019;60: 3–12.
24. Kordas K, Siegel EH, Olney DK, Katz J, Tielsch JM, Chwaya HM, et al. Maternal reports of sleep in 6–18 month-old infants from Nepal and Zanzibar: association with iron deficiency anemia and stunting. Early Hum Dev. 2008;84(6):389–98.
25. Leitner Y, Bloch AM, Sadeh A, et al. Sleep-wake patterns in children with intrauterine growth retardation. J Child Neurol. 2002;17(12):872–6.
26. Copes K, Rosentswieg J. The effects of sleep deprivation upon motor performance of ninthgrade students. J Sports Med Phys Fitness. 1972;12(1):47–53.
27. Berger SE, Scher A. Naps improve new walkers' locomotor problem solving. J Exp Child Psychol. 2017;162:292–300.
28. Stores G. Multifactorial influences, including comorbidities, contributing to sleep disturbance in children with a neurodevelopmental disorder. CNS Neurosci Ther. 2016;22(11):875–9.
29. Stores G. Sleep disturbance in specific neurodevelopmental disorders. Sleep and its disorders in children and adolescents with a neurodevelopmental disorder: a review and clinical guide. Cambridge: Cambridge University Press; 2014.
30. Mindell JA, Owens JA. A clinical guide to pediatric sleep: diagnosis and management of sleep problems. Lippincott Williams & Wilkins; 2015.
31. Melke J, Goubran Botros H, Chaste P, Betancur C, Nygren G, Anckarsäter H, et al. Abnormal mela-

tonin synthesis in autism spectrum disorders. Mol Psychiatry. 2008;13(1):90–8.
32. Mulder EJ, Anderson GM, Kemperman RF, Oosterloo-Duinkerken A, Minderaa RB, Kema IPJN. Urinary excretion of 5-hydroxyindoleacetic acid, serotonin and 6-sulphatoxymelatonin in normoserotonemic and hyperserotonemic autistic individuals. Neuropsychobiology. 2010;61(1):27–32.
33. Veatch OJ, Pendergast JS, Allen MJ, Leu RM, Johnson CH, Elsea SH, et al. Genetic variation in melatonin pathway enzymes in children with autism spectrum disorder and comorbid sleep onset delay. J Autism Dev Disord. 2015;45:100–10.
34. Souders MC, Mason TB, Valladares O, Bucan M, Levy SE, Mandell DS, et al. Sleep behaviors and sleep quality in children with autism spectrum disorders. Sleep. 2009;32(12):1566–78.
35. Reynolds S, Lane SJ, Thacker LJOO. Sensory processing, physiological stress, and sleep behaviors in children with and without autism spectrum disorders. OTJR. 2012;32(1):246–57.
36. Bonnet MH, Arand DL. Hyperarousal and insomnia: state of the science. Sleep Med Rev. 2010;14(1):9–15.
37. Wimpory D, Nicholas B, Nash S. Social timing, clock genes and autism: a new hypothesis. J Intellect Disabil Res. 2002;46(4):352–8.
38. Veatch OJ, Keenan BT, Gehrman PR, Malow BA, Pack AI. Pleiotropic genetic effects influencing sleep and neurological disorders. Lancet Neurol. 2017;16(2):158–70.
39. Veatch O, Veenstra-VanderWeele J, Potter M, Pericak-Vance M, Haines JJ. Genetically meaningful phenotypic subgroups in autism spectrum disorders. Genes Brain Behav. 2014;13(3):276–85.
40. Gregory AM, Sadeh A. Annual research review: sleep problems in childhood psychiatric disorders–a review of the latest science. J Child Psycol Psychiatr. 2016;57(3):296–317.
41. Stores GJ, Sleep and its disorders in children with a neurodevelopmental disorder: A review and clinical guide. 2014: 61–78.
42. Stores GJ. Sleep disturbance in childhood epilepsy: clinical implications, assessment and treatment. Arch Dis Child. 2013;98(7):548–51.
43. Herman JH, Sheldon SH. Pharmacology of sleep disorders in children. In: Principles and practice of pediatric sleep medicine. Philadelphia: Elsevier Inc; 2005. p. 327–38.
44. Hoban TF, Chervin RD. Sleep-related breathing disorders of childhood: description and clinical picture, diagnosis, and treatment approaches. Sleep Med Clin. 2007;2(3):445–62.
45. Foley LS, Maddison R, Jiang Y, Marsh S, Olds T, Ridley KJP. Presleep activities and time of sleep onset in children. Pediatrics. 2013;131(2):276–82.
46. Engelhardt CR, Mazurek MO, Sohl KJP. Media use and sleep among boys with autism spectrum disorder, ADHD, or typical development. Pediatrics. 2013;132(6):1081–9.
47. Seo WS. An update on the cause and treatment of sleep disturbance in children and adolescents with autism spectrum disorder. Yeungnam Univ J Med. 2021;38(4):275–81.
48. Stores G, Stores R. Sleep disorders and their clinical significance in children with Down syndrome. Dev Med Child Neurol. 2013;55(2):126–30.
49. Waddington H, McLay L, Woods L, Whitehouse AJO. Child and family characteristics associated with sleep disturbance in children with autism spectrum disorder. J Autism Dev Disord. 2020;50:4121–32.
50. Kamara D, Beauchaine TP. A review of sleep disturbances among infants and children with neurodevelopmental disorders. Rev J Autism Dev Disord. 2020;7:278–94.
51. Esbensen AJ, Schwichtenberg AJ. Sleep in neurodevelopmental disorders. Int Rev Res Dev Disabil. 2016;51:153–91.
52. Devnani PA, Hegde AU. Autism and sleep disorders. J Pediatr Neurosci. 2015;10(4):304–7.
53. Shelton AR, Malow B. Neurodevelopmental disorders common presenting with sleep disturbances. Neurotherapeutics. 2021;18:156–69.
54. Shelton AR, Malow B. Sleep disturbances in neurodevelopmental disorders. Curr Psychiatry Rep. 2016;18:6.
55. Jan JE, Owens JA, Weiss MD, Johnson KP, Wasdell MB, Freeman RD, et al. Sleep hygiene for children with neurodevelopmental disabilities. Pediatrics. 2008;122(6):1343–50.
56. Primeau M, Gershon A, Talbot L, Cotto I, Lotspeich L, Hardan A, et al. Individuals with autism Spectrum disorders have equal success rate but require longer periods of systematic desensitization than control patients to complete ambulatory polysomnography. JCSM. 2016;12(3):357–62.
57. Goldman SE, Bichell TJ, Surdyka K, Malow BA. Sleep in children and adolescents with Angelman syndrome: association with parent sleep and stress. JIDR. 2012;56(6):600–8.
58. Kamara D, Beauchaine TP. A review of sleep disturbances among infants and children with neurodevelopmental disorders. Rev J Autism Dev Disord. 2020;7(3):278–94.
59. Ashworth A, Hill CM, Karmiloff-Smith A, Dimitriou D. Cross syndrome comparison of sleep problems in children with down syndrome and Williams syndrome. Res Dev Disabil. 2013;34(5):1572–80.
60. Simard-Tremblay E, Constantin E, Gruber R, Brouillette RT, Shevell M. Sleep in children with cerebral palsy: a review. J Child Neurol. 2011;26(10):1303–10.
61. Maas AP, Didden R, Korzilius H, Braam W, Collin P, Smits MG, et al. Psychometric properties of a sleep questionnaire for use in individuals with intellectual disabilities. Res Dev Disabil. 2011;32(6): 2467–79.
62. Simonds JF, Parraga H. Prevalence of sleep disorders and sleep behaviors in children and adolescents. J Am Acad Child Psychiatry. 1982;21(4):383–8.

63. Poisson A, Nicolas A, Cochat P, Sanlaville D, Rigard C, de Leersnyder H, et al. Behavioral disturbance and treatment strategies in Smith-Magenis syndrome. Orphanet J Rare Dis. 2015;10:111.
64. Grigg-Damberger M, Ralls F. Treatment strategies for complex behavioral insomnia in children with neurodevelopmental disorders. Curr Opin Pulm Med. 2013;19(6):616–25.
65. Minshawi NF, Hurwitz S, Fodstad JC, Biebl S, Morriss DH, McDougle CJ. The association between self-injurious behaviors and autism spectrum disorders. Psychol Res Behav Manag. 2014;7:125–36.
66. Corkum P, Davidson FD, Tan-MacNeill K, Weiss SK. Sleep in children with neurodevelopmental disorders: a focus on insomnia in children with ADHD and ASD. Sleep Med Clin. 2014;9(2):149–68.
67. Ma ZR, Shi LJ, Deng MH. Efficacy of cognitive behavioral therapy in children and adolescents with insomnia: a systematic review and meta-analysis. Braz J Med Biol Res. 2018;51(6):e7070.
68. Pelayo R, Yuen K. Pediatric sleep pharmacology. Child Adolesc Psychiatr Clin N Am. 2012;21(4):861–83.
69. Heussler H, Chan P, Price AM, Waters K, Davey MJ, Hiscock H. Pharmacological and non-pharmacological management of sleep disturbance in children: an Australian Paediatric research network survey. Sleep Med. 2013;14(2):189–94.
70. Abdelgadir IS, Gordon MA, Akobeng AK. Melatonin for the management of sleep problems in children with neurodevelopmental disorders: a systematic review and meta-analysis. Arch Dis Child. 2018;103(12):1155–62.
71. Miyamoto A, Fukuda I, Tanaka H, Oka R, Araki A, Cho K. [Treatment with ramelteon for sleep disturbance in severely disabled children and young adults]. No to hattatsu = Brain and development 2013;45(6):440–444.
72. Schnoes CJ, Kuhn BR, Workman EF, Ellis CR. Pediatric prescribing practices for clonidine and other pharmacologic agents for children with sleep disturbance. Clin Pediatr. 2006;45(3):229–38.
73. Ingrassia A, Turk J. The use of clonidine for severe and intractable sleep problems in children with neurodevelopmental disorders--a case series. Eur Child Adolesc Psychiatry. 2005;14(1):34–40.
74. Rugino TA. Effect on primary sleep disorders when children with ADHD are administered Guanfacine extended release. J Atten Disord. 2018;22(1):14–24.
75. Arens R, Wright B, Elliott J, Zhao H, Wang PP, Brown LW, et al. Periodic limb movement in sleep in children with Williams syndrome. J Pediatr. 1998;133(5):670–4.
76. Robinson AA, Malow BA. Gabapentin shows promise in treating refractory insomnia in children. J Child Neurol. 2013;28(12):1618–21.
77. Bruni O, Angriman M, Melegari MG, Ferri R. Pharmacotherapeutic management of sleep disorders in children with neurodevelopmental disorders. Expert Opin Pharmacother. 2019;20(18): 2257–71.

Family Health and Neurodevelopment

Peter Schwandt and Gerda-Maria Haas

Abstract

Objective
We examined whether associations exist between maternal consumption of vitamins, whose effects on the neurodevelopment of fetus and embryos are known, and the vitamin status of offspring.

Methods
Out of the PEP Study, we invited 687 biological child–parent pairs who fulfilled the conditions for family history, physical examination, blood sampling, a 7-day weighed dietary protocol, and general health advice. According to the evaluation of macronutrients from the dietary protocols, we now evaluated the protocols for vitamins.

Thereby, we calculated the means of daily consumption (μg) of Folic Acid, β-carotene, Vitamin A, Vitamin B6, Vitamin B12, Vitamin C, Vitamin D, and Vitamin E for mothers, fathers, daughters, and sons ($p < 0.05$). Compared with mothers, fathers had higher consumption of vitamins than mothers. The goals of the German Society of Nutrition were better reached by parents than by children. Using General Estimating Equations, we calculated whether there were direct associations between the unsatisfied vitamin status of parents and the unsatisfied vitamin status of parents. Significant Odds Ratios were found for vitamins B6, C, D, and E, with 13.5 (CI 95% 5.4–33.5), 6.9 (CI 95% 2.9–16.3), 4.5 (CI 95% 2.1–9.3), and 11.4 (CI 95% 4.2–30.8), with $p < 0.000$.

P. Schwandt (✉)
Head, Arteriosklerose-Praeventions-Institut, München, Germany

G.-M. Haas
Atherosclerosis Prevention Institute, Muenchen, Germany

1 Introduction

Some vitamins, such as Vitamin B12, are known as important for cell division and differentiation, or necessary for the development of the central nervous system when mothers consume it during pregnancy. Studies found that Vitamin B12 influences neurodevelopment in newborn children as well as in chidhood [1]. Another study found this relationship in school-children, too [2]. Other studies examined the relationship of Vitamin D levels, e.g., with neurodevelopment in children, too [2, 3]. The influence of folic acid, e.g., for neurodevelopment is well known. Consumption of vitamins in daily nutrition is an important fator in developing a healthy brain early in life, with long-lasting effects. It often irreversibly influences an individual's cognitive development and lifelong mental health [4]. The aim of our study

© The Author(s), under exclusive license to Springer Nature Singapore Pte Ltd. 2024

R. Kelishadi (ed.), *Nature, Nurture, and Neurodevelopment*,
https://doi.org/10.1007/978-981-97-7433-3_9

was to examine whether these positive effects of vitamins for neurodevelopment are passed on from mothers to their school-aged children, and whether a deficiency of some vitamins was passed on from parents to children or not. Therefore, we observed in 687 child–parent pairs the nutrition behavior between school-aged children and their families out of the big PEP-Family Heart Study [5–7].

2 Methods

As described before, the study design of the PEP Family Health never had been changed within all substudies over time [5].

Here, we only report the first year. Whereas former manuscripts described the macronutrients and their relationship to Cardiovascular Risk Factors, here we examined for our family observation manuscript only β-carotene, Vitamin A, Vitamin B6, Vitamin B12, Vitamin C, Vitamin D, and Vitamin E, as well as Folic Acid, as recommended for pregnant women [4]. The aim was to examine whether the mothers, who most times are responsible for family nutrition, pass on these recommendations to their children.

2.1 Study Population

We invited 2690 parents and their children with complete risk factor profiles and lifestyle data [7] to join the 2-year follow-up study described here voluntarily. The inclusion criteria included 7-day weighed dietary protocols for mothers as well as for their children. Therefore, we report here on 575 parents and 411 biological children who successfully completed the 2-year follow-up study. These participants consisted of 254 fathers (mean age 39.2 ± 5.2 years) and 321 mothers (36.4 ± 4.1 years).

2.2 Exclusion Criteria

Here, we accepted only German participants to avoid ethnic bias. A total of 2.6% of the participants from 17 other or mixed ethnic groups were excluded. Criteria for noninclusion too were self-reported cardiovascular, metabolic, endocrine, and malignant disorders, extreme physical activity, special nutritional habits, and medication.

3 Measurements

Regularly trained research assistants performed all measurements as described previously [5–8]. Briefly, personal and familial medical history, socioeconomic data, medication, smoking habits, and leisure time physical activity (LTPA), based on metabolic equivalents (METs) according to Ainsworth and Ridley [9, 10], were assessed by interviewer-assisted questionnaires. Food intake was recorded by 7-day weighed dietary protocols. As described previously, all measurements were performed by trained research assistants [5–8]. Besides height, weight, waist circumference (WC), hip circumference (HC), skinfold thickness (SFT), systolic blood pressure (SBP), diastolic blood pressure (DBP), and fasting lipid and glucose concentrations were measured. Body mass index (BMI), SFT sum, waist-to-height ratio (WHtR), waist-to-hip ratio (WHR), non-high-density lipoprotein cholesterol (HDL-C), low-density lipoprotein cholesterol (LDL-C) to HDL-C ratio, triglyceride (TG) to HDL-C ratio, and percentage body fat (%BF) according to Slaughter equations from SFT were calculated.

4 Nutrition

All participants were trained to precisely document their daily intake of food and beverages over seven consecutive days, including a weekend. The participating families had to fill in their dietary record forms together with an accurately calibrated digital food scale (Soehnle combi plus, Nassau, Germany).

Most of the time, the mothers weighed the meals and wrote the protocols for all family members.

Completed dietary records were analyzed by trained dieticians and ecotrophologists using the computer program PRODI (version 4.5, Nutri Science, Freiburg, Germany), which includes the "Deutscher Lebensmittelschlüssel" supplemented with individual special items. The PRODI data were transferred into SPSS (version 15.0) using the American Standard Code for Information Interchange (ASCII) for documentation and calculation. Normal consumption of nutrients was defined according to the weight-, age-, gender-, and moderate physical activity-adjusted nutrient intakes in Germany, Austria, and Switzerland (D-A-CH) [11].

For children, we calculated the deviation below the median values as a limit according to the D-A-CH values, based on the values of the Deutsche Gesellschaft für Ernährung 2008 [11].

For this branch of the study, we only observed the first year of participation.

5 Results

Accordingly, for families with first graders, we found the means for parents and children in Table 1.

Tables 2–5 show the mean values of the examined vitamins for parents (Tables 2 and 3) and children (Tables 4 and 5).

We observed the following vitamins: Vitamin A, β-carotene, Vitamin B6, Vitamin B12, Vitamin C, Vitamin D, Vitamin E, and Folic Acid. Therefore, we followed the recommendations of the DGE as shown previously for the recommended daily intake for adults between 25 and 65 years.

Compared to mothers, most fathers consumed more vitamins than mothers.

Noticeable standard deviations were found for each of the vitamins compared to the mean values. The highest SD was observed for Vitamin D in daughters compared with the mean values of this group.

As breastfeeding's influence on the next 2 years of infant growth and neurodevelopment is described [12] by Agostini et al., we found evidence for the vitamins above and their positive effects on neurodevelopment, as well for the emotional development, and the grasping of the world through healthy eyes in many publications of the last years [13, 14].

Table 1 Characteristics of the participants

	Mothers	Fathers	Daughters	Sons
Mean age(y)	36.4(4.1)	39.2(5.1)*	6.8(1.8)	6.8(1.7)
Height (cm)	166.1(6.3)	180.2(6.8)*	125.1(12.0)	125.1(11.2)
Weight (kg)	66.6(11.1)	82.2(11.5)*	25.3(7.5)	25.3(7.1)
BMI (kg/m^2)	23.7(3.7)	25.3(3.2)*	15.8(2.0)	15.9(2.0)
Tobacco smoking	12.7%	26.0%*	0	0

$p > 0.05$.

Table 2 Vitamin status of the parents (Mothers n = 321)

Vitamins	Mean (g/day)	SD
Mothers		
B-carotene	2427	1869
Vitamin A	0.4	0.2
Vitamin B6	1.2	0.4
Vitamin B12	3.1	1.2
Vitamin C	89	45
Vitamin D	1.6	1.1
Vitamin E	7.6	3.8
Folic acid	192	106

Table 3 Vitamin status of the parents (Fathers n = 254)

Vitamins	Mean (g/day)	SD
Fathers		
B-carotene	2517	2018
Vitamin A	0.4	0.2
Vitamin B6	1.2	0.4
Vitamin B12	3.1	1.3
Vitamin C	90	48
Vitamin D	1.7	1.4
Vitamin E	7.7	3.0
Folic acid	193	108

Table 4 Vitamin status of the parents (Daughters n = 216)

Vitamins	Mean (g/day)	
Daughters		
B-carotene	3016	1917
Vitamin A	0.6	0.7
Vitamin B6	1.6	0.5
Vitamin B12	5.1	2.8
Vitamin C	105	95
Vitamin D	4.3	3.8
Vitamin E	10.9	4.3
Folic acid	261	214

Table 5 Vitamin status of the parents (Sons n = 195)

Vitamins	Mean (g/day)	
Sons		
B-carotene	3680	3104
Vitamin A	0.6	0.4
Vitamin B6	1.6	0.4
Vitamin B12	5.0	2.4
Vitamin C	100	47
Vitamin D	4.0	3.5
Vitamin E	10.8	3.6
Folic acid	249	100

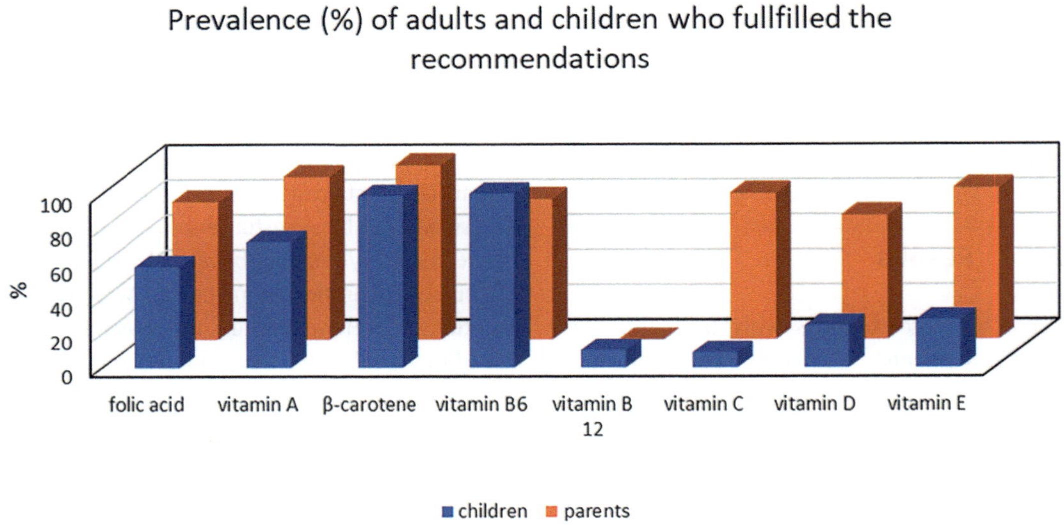

Fig. 1 Prevalences of parents and children following the recommendation of daily consumption of vitamins

For adults, the German Society of Nutrition recommended [9] fixed values for vitamin consumption depending on age. For children, the Society was oriented toward the medians of consumption depending on age as well (Fig. 1).

6 Prevalences

Folic Acid recommendations were achieved by double percent of children compared to parents, but Vitamin A was three-fold higher in parents than in children. Only the β-carotene recommendation was fulfilled in both parents and children. The recommendations for daily consumption of Vitamin B6 (40-fold), Vitamin B12, as well as Vitamin C (ten-fold), Vitamin D (three-fold), and Vitamin E (3.5-fold) were better fulfilled in parents than in children.

7 General Estimating Equations (GEE)

According to our findings for the macronutrients, for those who did not reach the goals, we calculated using GEE whether there are comparable findings for vitamin consumption as well. This means whether these parents (both fathers and mothers) who fail to meet the goals pass on this deficiency to their offspring.

Table 6 Probability for children of parents with vitamin deficiency to also fail the recommended vitamin consumption

Vitamin	P	Odds ratio	Lower level	Upper level
Vitamin B6	P < 0.000	13.5	5.4	33.5
Vitamin E	P < 0.000	11.4	4.2	30.8
Vitamin C	P < 0.000	6.9	2.9	16.3
Vitamin D	P < 0.000	4.5	2.1	9.3

These relationships, calculated with Exp B (Odds), did not find these equivalents for deficiency in Vitamin A and Vitamin B12 but only with a significant ($p < 0.001$) weak equivalent <1.

The results were demonstrated in Table 6.

The risk of a lack of Vitamin B6 was observed to be 13.5-fold higher, 11.4-fold higher for Vitamin E, 6.9-fold higher for Vitamin C, and 4.5-fold higher for Vitamin D in children whose parents also had this deficiency.

As shown before no deficiency of β-carotene was found in both, in children as well as in parents.

8 Conclusion

It is necessary to make more known the importance of Vitamin consumption during pregnancy as well for the growth of the brain and emotions of newborns, infants, and school-aged children, too. Therefore, we need material for pregnant and young mothers to inform them about the deficits of important vitamins for their children and how to find all these vitamins in their daily nutrition as follows:

Vitamin A [4, 15]

The main consequences of Vitamin A deficiency are night blindness, anemia, and immunodeficiency.

Retinoids are essential for sight, normal embryonic development, and the control of cell growth, differentiation, and death. It may cause teratogenic effects in body organs during pregnancy.

Recommendations should be:

Carots, apricots, green cabbage, and spinach.

B-carotene [4]

Is a pre-step for Vitamin A.

Nutrition recommendations should be:

Grapefruit red, pepper red, savoy cabbage, and lamb's lettuce.

Vitamin B6 [4, 16]

For immune function, gene transcription, as well as for gene expression, B6 is essential and important for brain glucose regulation.

Recoommendations should be:

Fat fish (e.g., mackerel), natural mushrooms, and yolk.

Vitamin B12 [1–3]

Vitamin B12 plays an important role in Central Nervous System metabolism. Deficiency-related syndromes include myelopathy, neuropathy, neuropsychiatric disorders, and seldom optic nerve atrophy.

Recommendations should be:

Animal foods, lax, and cow milk.

Vitamin C [4, 18]

A lot of enzymatic reactions are influenced by Vitamin C, including redox balance in the brain, modulating the cholinergic, catecholaminergic, and glutamatergic systems, generally developing neurons through their maturation and differentiation, and forming myelin. In the developing brain, neuronal density and maturation are affected by a lack of Vitamin C, which is related to brain volume. Vitamin C deficiency influences folic acid metabolism.

Recommendations should be:

Fruits, juice of natural fruits, peppers, parsley, sea buckthorn, haws, and black currants.

Vitamin D [6, 16, 17]

Vitamin D (1,25-dihydroxycholecalciferol) is related to nerve growth. It nurtures neurite growth and suppresses neuronal apoptosis in the hippocampus. During the prenatal period, Vitamin D deficiency inhibits the expression of certain genes involved in brain and cerebellar growth.

Recommendations should be:

Fat fishes, macarel, yolk, and natural mushrooms.

Vitamin E [4, 18]

Vitamin E deficiency is associated with cerebellar ataxia, posterior cord medullary damage, and peripheral neuropathy.

Recommendations should be:

Plant Oil, nuts, sunflower seeds, and green cabbage.

Folic Acid (B9) [4, 19, 20]

Vitamin B9 is important for cell division. It is essential for growth and for the normal development of the fetal spine, brain, and skull. Especially during the first four weeks of gestation, folic acid is essential for growth and for the normal development of the fetal spine, brain, and skull.

Recommendations should be:

Greens, spinach, green salads, tomatoes, lentils, pies, nuts, oranges, potatoes, and eggs.

These recommendations are valid not only for maternal nutrition during pregnancy but also for growth in the following years of infancy and school-children.

So our advice is:

At the begin of pregnancy women should get written material for breastfeeding as well as dietary advice for neurodevelopment for their children with their motherhood passports.

References

1. Maureen M. Effects of vitamin B12 and folate deficiency on brain developmentin children. Food Nutr Bull. 2008;29(2 Suppl):S126–31. NIH Public Access Author Manuscript Food Nutr Bull manuscript; available in PMC 2011 July 18. Published in final edited form
2. Winje BA, et al. Does early vitamin B12 supplementation improve neurodevelopment and cognitive function in childhood and into school age: a study protocol for extended follow-ups from randomised controlled trials in India and Tanzania. BMJ Open. 2018;8:e018962. https://doi.org/10.1136/bmjopen-2017-018962.
3. StrandI TA, et al. Effects of vitamin B12 supplementation on neurodevelopment and growth in Nepalese Infants: a randomized controlled trial. PLOS Medicine. 2020;17:e1003430.
4. Heland S, Fields N, Ellery SJ. The role of nutrients in human neurodevelopment and their potential to prevent neurodevelopmental adversity. Front Nutr. 2022;9:992120. https://doi.org/10.3389/fnut.2022.992120.
5. Schwandt P, Geiss HC, Ritter MM, et al. The prevention education program (PEP). A prospective study of the efficacy of family-oriented life style modification in the reduction of cardiovascular risk and disease: design and baseline data. J Clin Epidemiol. 1999;52:791–800.
6. Schwandt P, Bischoff-Ferrari HA, Staehelin HB, Haas GM. Cardiovascular risk screening in schoolchildren predicts risk in parents. Atherosclerosis. 2009;205:626–31.
7. Schwandt P, Bertsch T, Haas G-M. Sustained lifestyle advice and cardiovascular risk factors in 687 biological child–parent pairs: The PEP Family Heart Study. Atherosclerosis. 2011;219(2):937–45. published by Elsevier Ireland. https://doi.org/10.1016/j.atherosclerosis.2011.09.032.
8. Tröbs M, Renner T, Scherer G, et al. Nutrition, antioxidants, and risk factor profile of nonsmokers, passive smokers and smokers of the prevention education program (PEP) in Nuremberg. Germany Prev Med. 2002;34:600–7.
9. Ridley K, Ainsworth BE, & Olds TS. Development of a Compendium of Energy Expenditures for Youth. International Journal of Behavioral Nutrition & Physical Activity. 2008;5:1–8. https://doi.org/10.1186/1479-5868-5-45.
10. Development of a compendium of energy expenditures for youth Kate Ridley 1 , Barbara E Ainsworth, Tim S Olds Int J Behav Nutr Phys Act. 2008;5:45. https://doi.org/10.1186/1479-5868-5-45.
11. Ernährungsbericht 2008 Editor: Deutsche Gesellschaft für Ernährung. ISBN: 978–3.
12. Agostoni C, et al. The long-term effects of dietary nutrient intakes during the first 2 years of life in healthy infants from developed countries: an umbrella review. Adv Nutr. 2019;10(3):489–501.
13. Lockyer F, McCan S, Moore SE. Micronutrients and infant neurodevelopmental outcomes: a systematic review. Nutrients. 2021;13(11):3848. https://doi.org/10.3390/nu13113848.
14. Yang T, et al. Vitamin A status is more commonly associated with symptoms and neurodevelopment in boys with autism spectrum disorders-a multicenter

study in China. Front Nutr. 2022;9:851980. https://doi.org/10.3389/fnut.2022.851980. eCollection 2022
15. Field DT. High-dose vitamin B6 supplementation reduces anxiety and strengthens visual surround suppression. Human Psycopharmacol. 2022;37(6):e2852.
16. Moritz B, Schmitz AE, Rodrigues ALS, Dafre AL, Cunha MP. The role of vitamin C in stress-related disorders. J Nutr Biochem. 2020;85:108459. https://doi.org/10.1016/j.jnutbio.2020.108459.
17. Navale SS, Mulugeta A, Zhou A, Llewellyn DJ, Hyppönen E. Vitamin D and brain health: an observational and Mendelian randomization study. Am J Clin Nutr. 2022;116(2):531–40. https://doi.org/10.1093/ajcn/nqac107.
18. Traber MG. Vitamin E: necessary nutrient for neural development and cognitive function. Proc Nutr Soc. 2021;80(3):319–26. Published online by Cambridge University Press: 26 April 2021
19. Helene MN. Effect of continued folic acid supplementation beyond the first trimester of pregnancy on cognitive performance in the child: a follow-up study from a randomized controlled trial (FASSTT offspring trial). BMC Med. 2019;17:196.
20. Pujol C, et al. Implication of folate deficiency in CYP2U1 loss of function. J Exp Med. 2021;218(11):e20210846. https://doi.org/10.1084/jem.20210846.

Early Screening of Neurodevelopment

Maryam Bemanalizadeh, Negin Badihian, Seyede Shahrbanoo Daniali, and Omid Yaghini

Abstract

Early screening of neurodevelopment in children enables primary healthcare providers to make earlier referrals to interventional programs. Evidence shows that children who are assigned to therapeutic services are more likely to achieve their potential development. In this chapter, we provide an overview of the developmental milestones and the latest classification of developmental delays (DDs) and neurodevelopmental disorders (NDDs) in children. In addition, the psychometric properties of the most common tools for screening development in children, including the Ages and Stages Questionnaire (ASQ), ASQ: Social-Emotional (ASQ: SE), Parents' Evaluation of Developmental Status (PEDS), Bayley Scales of Infant and Toddler Development (BSID), and Denver Developmental Screening Test (DDST), are presented in this chapter. Along with these standard screening tools, we also discuss groundbreaking research discovering some novel biomarkers or methods for a more accurate diagnosis of DDs or NDDs.

Keywords

Development · Neurodevelopmental disorders · Developmental delays · Motor development · Language development · Cognition · Social-emotional development · Screening · Diagnosis · Surveillance

M. Bemanalizadeh · S. S. Daniali
Child Growth and Development Research Center, Research Institute for Primordial Prevention of Non-Communicable Disease, Isfahan University of Medical Sciences, Isfahan, Iran
e-mail: m.bemanalizadeh@mui.ac.ir

N. Badihian
Department of Neurology, Mayo Clinic, Rochester, MN, USA

O. Yaghini (✉)
Child Growth and Development Research Center, Research Institute for Primordial Prevention of Non-Communicable Disease, Isfahan University of Medical Sciences, Isfahan, Iran

Department of Pediatric Neurology, Imam Hosein Hospital, Isfahan University of Medical Sciences, Isfahan, Iran

1 Background

The first 5 years of life are crucial for cognitive, language, social, and motor development [1]. Although the brain's ability to adapt enables us to learn new skills throughout life, healthy development in the early years of life is vital for later academic achievements, economic growth, behavioral and social communication, good citizenship, lifetime health, and successful parenting of the coming generations. Early experiences shape the brain's architecture and create foundations for identity formation and lifelong functioning [2–4]. Over one thousand trillion synapses

© The Author(s), under exclusive license to Springer Nature Singapore Pte Ltd. 2024
R. Kelishadi (ed.), *Nature, Nurture, and Neurodevelopment*,
https://doi.org/10.1007/978-981-97-7433-3_10

exist in the central nervous system (CNS), and more than one million new synaptic connections develop each second during the early years of life [5]. The development and maturity of the brain and CNS require a more precisely organized series of events and processes than any other system, making it especially susceptible to prenatal and postnatal environmental events and exposures [6].

Neurodevelopmental disorders (NDDs) and developmental delays (DDs) emerge in early childhood: NDDs present with cognitive impairments and abnormalities in communication, behavior, and/or motor skills and are consequences of abnormal brain development, while DDs are defined as the failure to reach developmental milestones in children under 5 years. NDDs can be associated with lifetime impairments with poor prognosis, lower quality of life for children and their families, and higher medical service costs [7–10]. Although interventions during later years can improve these conditions to a considerable degree, they will be less efficient and require significantly more effort. Although the prevalence of DDs and NDDs varies between populations and depends on the definitions and data sets that have been analyzed, the growing number of new cases is remarkable [11]. Approximately 10% of children have one or more DDs or NDDs, and many more are believed to require support in school or other situations in some critical areas of functioning [12].

Early diagnosis of children with DDs or NDDs, initiating appropriate intervention (e.g., speech/language therapy, physical therapy, occupational therapy, or special education services, etc.), and follow-up are acknowledged as vital aspects of excellent community care. Without them, early difficulties may remain or worsen over time. The prosperous detection of children with a DD or NDD needs two primary steps: (i) screening for the risk of DDs or NDDs using a standardized screening instrument and (ii) a full diagnostic assessment of children whose development deviates from the same-age norms [1, 13]. Early interventional programs can plan for educational, preventive, or therapeutic services for children with risk factors [2, 4, 14, 15] and can improve long-term outcomes in a wide range of areas (Fig. 1) [16–20]. Focused early intervention programs may be particularly beneficial for children residing in low-resource or low-educational status [16, 21].

2 Main Developmental Domains and Red Flags

Developmental delay can affect a single stream of development (single domain) or two or more streams of development (global developmental delay) [22, 23]. Conceptually, the domains of development can be classified into four broad categories:

1. Gross and fine motor abilities are the two components of motor development. Gross motor abilities are defined as the control of vast groups of muscles involved in walking, sitting, or shifting from one posture to another, while fine motor abilities are defined as the manipulation of objects with the hands in order to eat, draw, write, play, etc. Children move through motor milestones in a systematic manner, acquiring these abilities in a clear and sequential process. Motor delay is characterized by a prominent delay in motor skills without other developmental delays.
2. Language performances, which include articulation, receptive and expressive language skills, and the use of nonverbal symbols, constitute a major stream of development, resulting from the interaction between innate communication capacities and environmental factors. A developmental language disorder, developmental dysphasia, or specific language impairment is classified as any significant delay in language or speech skills without a delay in other developmental domains.
3. Adaptive or cognitive development is the child's capacity to solve problems using intuition, perception, and verbal and nonverbal reasoning. Moreover, it includes not only the ability to learn and comprehend but also the capacity to remember and apply this information when necessary.

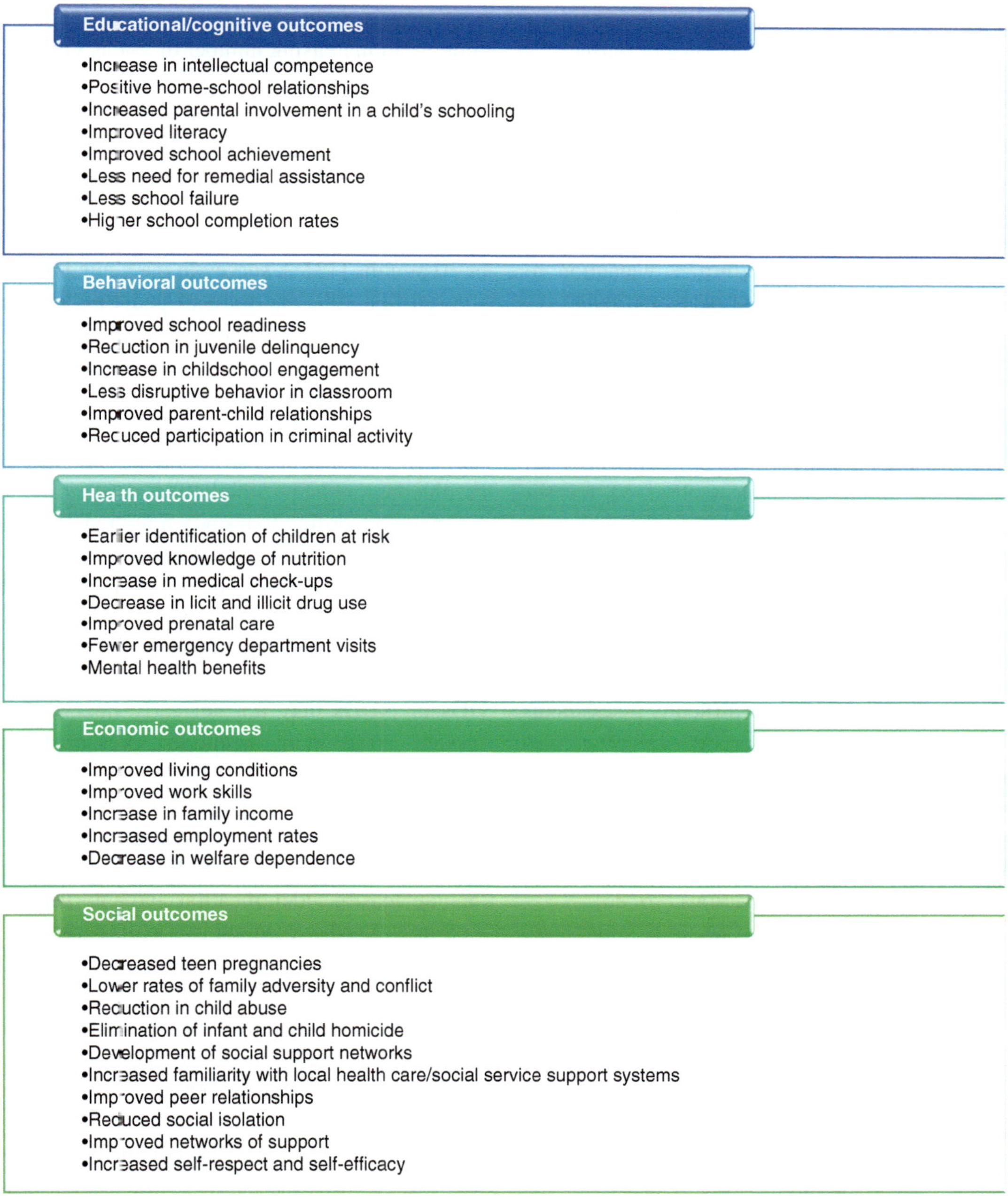

Fig. 1 An overview of the benefits of early identification and intervention of DDs and NDDs in various aspects

4. Personal-social-emotional or behavioral development includes the child's interactions, as demonstrated by making new relationships and keeping them, and reacting to the presence of others. Over time, psychosocial delay manifests as behavioral abnormalities that differ from typical behavioral responses in their quantity, severity, nature, and duration. Personal development entails the formation of self-help skills in a variety of daily activities, including feeding, dressing, and toileting [23–25].

In developmental milestones, clinicians may identify red flags that lead to concern, further surveillance, or referral. The absence of some developmental milestones or the reporting of specific behaviors in a child by the parent may suggest a disorder that is more likely to be persistent or require earlier intervention. Children with early motor disabilities or hypotonia of unknown etiology may benefit from further evaluation for conditions such as cerebral palsy, muscular dystrophy, or other neuromuscular disorders. The latest American Academy of Pediatrics (AAP) recommendation for the early detection of motor delay presents a useful approach for motor evaluation at specific time points. Based on this approach, when hypotonia is observed, measuring creatine kinase and thyroid function tests are recommended. Moreover, in cases of persistent, unexplained hypertonia, brain magnetic resonance imaging should be considered. Children who manifest red flags in social communication should be recommended for autism spectrum disorder or language difficulty assessments. Children with receptive or expressive language delays may benefit from a comprehensive assessment and intervention by a speech/language pathologist. Children whose developmental delay cannot be explained by medical history may take advantage of genetic evaluations [16, 26].

3 Classification of Neurodevelopmental Disorders

Infants, children, and adolescents vary in their major clinical presentation of atypical development. Figure 2 shows the Diagnostic and Statistical Manual of Mental Disorders (DSM-5) classification of NDDs [7].

4 Early Screening Across Different Countries

To satisfy the requirements of children during the most crucial stage of their development, numerous nations have built programs and services to alleviate disabilities [27]. Early intervention and early childhood special education support an increasing number of young children with NDDs and DDs and their families [21, 28]. The Canadian Paediatric Society endorses developmental screening of all children at 18 months using a standard screening instrument [29], while the American Academy of Pediatrics supports developmental screening at 9, 18, and 24 or 30 months [13]. A survey in the United States showed that still, one-third of pediatricians are not following the standardized screening program due to several causes, such as time limitations, inadequate reimbursement, and lack of treatment options for positive screening results [30]. In Canada, socioeconomic and demographic factors had a small effect on the time to diagnosis according to a study that modeled this issue. This study reveals that health system features and regional differences in wait times are the most important determinants of time to diagnosis [31]. The delay in diagnosis restricts accessibility to early treatment, as children with DDs or NDDs may not get early intervention services until their symptoms present more severely or when they have a definite diagnosis [32]. There is evidence that protective variables such as social care, welfare assistance, and community involvement for children at risk could include feasible identification and intervention measures in the community with parents to reduce the risk of child developmental and behavioral problems [33].

In Singapore, children are scheduled to be visited by trained nurses or doctors for the Denver Developmental Screening Test at specific ages for six rounds between the ages of 1 month and 4–6 years [34]. In some countries, such as Iran and Norway, the public health system provides free medical facilities for all children, and close to 100% of parents with infants and young children come regularly to local primary healthcare centers or well-baby clinics from birth to 5 years of age for weight control, vaccination, and developmental screening of their infant [35, 36]. There is scarce information on NDDs from lower- and middle-income countries. A review by Samms-Vaughan et al. indicated that the mean age of parental concern regarding their children's symptoms has been between 21 and

Fig. 2 Neurodevelopmental disorders classification based on the fifth edition of diagnostic and statistical manual of mental disorders (DSM-5) [7]

24 months, and the mean age of diagnosis for autism spectrum disorders among their children has been approximately 45 to 57 months. The high delay in diagnosis across vastly different lower- and middle-income countries suggests limited resources and a lack of strong early identification systems for NDDs in these countries [37]. In a multinational study on 16 developing countries, it has been shown that the mean frequency of children who screened positive for at least one DD was 20.4%. In general, younger children and countries with lower living standards were more likely to be screened as having some type of disability [38].

5 Most Common Tools for Screening and Their Psychometric Properties

Accurate diagnosis and treatment, policy-making, and ultimately service delivery depend on the questionnaire's ability to yield accurate information and detect individuals with NDDs or

DDs. Here, we present some of the most common tools that are used for neurodevelopment screening in the early years of life.

5.1 Ages and Stages Questionnaire (ASQ)

The Ages and Stages Questionnaire (ASQ) is a standardized parent-reported questionnaire that may be widely applied as a general developmental screening tool for children. The ASQ was designed by J. Squires et al. in 1995 [39], and the third edition was developed in 2009 [40]. It approximately takes 12–18 minutes to complete and consists of 21 age-specific intervals covering 2 months to 66 months of age, each with 30 questions in five domains. These five domains are as follows: (i) personal social, (ii) gross motor, (iii) fine motor, (iv) problem-solving, and (v) communication, which can be answered with "yes" (10 points), "sometimes" (5 points), or "not yet" (0 points) answers, along with some open-ended questions at the end of the questionnaire. Each domain with six questions results in a domain-specific score that is compared with determined age-specific cut-off points at one and two standard deviations (SD) that are used to recognize high-risk children in terms of developmental delay. If any domain score comes below the 2SD cut-off, additional evaluation is recommended. If the score in any domain is between the 1SD and 2SD cut-off points, it is suggested to provide learning activities and closely observe the child's development [1]. In the majority of cases, these surveys reliably detect young children who require an additional assessment to demonstrate their eligibility for early intervention programs [2]. The ASQ has been confirmed to be a standard screening tool with acceptable psychometric parameters, including test–retest reliability of 92%, sensitivity of 87.4%, and specificity of 95.7%. Its validity has also been evaluated across various cultures and populations worldwide [41–47]. The ASQ is accurate, practical, and widely used in multiple countries. There is a growing number of studies on its psychometric properties in a variety of cultural backgrounds (e.g., Australia, Brazil, Canada, Chile, China, Denmark, Ecuador, France, Ghana, India, Iran, Korea, Lebanon, the Netherlands, Norway, Republic of Macedonia, Spain, Taiwan, Argentina, Indonesia, South Africa, Zambia, Thailand, Turkey, etc.). The ASQ has been translated into multiple languages, including but not limited to English, Arabic, Chinese, French, Spanish, Vietnamese, Danish, Hmong, Italian, Japanese, Somali, Norwegian, Chilean, Brazilian, Turkish, Dutch, Persian, Hindi, Thai, and Korean [2, 41–44, 48–60]. Another benefit of the ASQ is its adaptability; it is useful in a wide range of contexts, including home, parenting programs, preschool programs, doctors' offices, health clinics, and early intervention units. This questionnaire can be completed either individually by parents/caregivers or with an expert's aid or even by a trained professional who knows the child well. Furthermore, the Joint Committee on Screening and Diagnosis of Autism suggests it for early identification of autism along with its application for general developmental screening and monitoring in office settings or primary care centers [61]. The ASQ has been extensively used for follow-up and assessment of preterm infants, at-risk newborns, and those who were born through assisted reproductive methods as well. As a parent-reported instrument, it provides a valuable platform for parents to discuss their concerns regarding their child's development in a systematic setting with a healthcare professional [62].

5.2 Ages and Stages Questionnaires: Social-Emotional (ASQ: SE)

ASQ: SE is focused only on social-emotional development in children. Accurate detection of behavior through the ASQ: SE provides the initial background for further steps of clinical evaluation, specific intervention, or close monitoring. ASQ: SE is a parent-completed tool that was developed to be used alone or in combination with the ASQ or other screening tools [63]. The ASQ: SE-2, the latest version of ASQ: SE, can be employed to screen children aged 1 to 72 months and has age-specific questions corresponding to 2, 6, 12, 18, 24, 30, 36, 48, and 60 months of age. It was developed to evaluate seven distinct scales of social-emotional development, including self-regulation, adaptive functioning, autonomy, compliance, affect, communication, and interaction with people. The number of questions varies between ages, and each question can be answered with "most of the time" (0 points), "sometimes" (5 points), and "rarely or never" (10 points). There are also some open-ended questions at the end of each questionnaire. A high score on the ASQ: SE shows delays in social-emotional development. ASQ: SE indicated a strong internal consistency of 0.82 and test–retest reliability of 94%, as well as a strong concurrent validity. Both strong psychometric properties and accessibility (e.g., quick completion) make the ASQ: SE an efficient option for researchers and clinicians performing large-scale studies or interventions [64]. The teacher-reported version of ASQ: SE also demonstrated promising findings regarding its use in school settings [65]. This questionnaire is available in English, Arabic, French, Spanish, Danish, Hmong, Somali, Turkish, Chinese, Brazilian, Swedish, Norwegian, and Colombian [48, 57, 66].

5.3 Parents' Evaluation of Developmental Status (PEDS)

Parents' Evaluation of Developmental Status (PEDS) is an accurate, evidence-based instrument for identifying and spotting developmental and behavioral problems in children from birth to 8 years of age. PEDS is a brief parent-based questionnaire consisting of 10n items: 8 yes/no questions and 2 open-ended questions. It takes 2–5 minutes to complete, making it feasible for use at the office, clinic, or home. PEDS has a sensitivity of 75% and a specificity of 74% [67]. It does not follow a numeric scoring but helps categorizing the child into one of three groups: low-, medium-, or high-risk. Children who are medium- or high-risk based on PEDS would be referred for further evaluations [23]. A previous study conducted in urban pediatric clinics showed that physicians were able to detect developmental problems more confidently and earlier when they used PEDS, and the effectiveness of the visits and subsequent care was enhanced [68].

5.4 Bayley Scales of Infant Development (BSID)

Bayley Scales of Infant and Toddler Development (BSID) is a comprehensive, structured assessment instrument for evaluating DDs in early childhood. In clinical and research settings, BSID has been utilized for several decades as an effective diagnostic for detecting early developmental delay [69]. It can also be used for preterm infant assessments [70]. Nancy Bayley released the first BSID in 1969. The first BSID evaluated motor and cognitive domains in infants aged 3 to 28 months. BSID second edition, which was published in 1993, added one more scale, the behavior rating scale, and expanded the age group from 1 to 42 months. Finally, BSID 3, which was published in 2006, was designed to evaluate development from 1 to 42 months of age in five domains: cognitive, motor, language, social-emotional, and adaptive behavior. BSID 4 was released in 2019 and is currently in use. BSID 4 consists of a cognitive scale (81 items), a language scale (42 items in the receptive and 37 items in the expressive domain), a motor scale (46 items in the fine motor and 58 items in the gross motor domain), and a social-emotional scale (which is derived from the Greenspan Chart and adaptive behavior scale that employs the Vineland behavior assessment system). The latest version of BSID is utilized for children aged 16 days to 42 months. Although BSID 4 is more time-saving and has greater clinical sensitivity and specificity when compared to BSID 3, it still takes about 30 to 70 minutes to complete. It should be mentioned that environmental, psychological, and physical testing settings have a substantial influence on BSID assessment scores, and trained professionals are required for its implementation [71].

5.5 Denver Developmental Screening Test (DDST)

The Denver Developmental Screening Test (DDST) is one of the most extensively used screening tools and has been administered among different populations. It can evaluate children between 0 and 6 years of age and takes about 20 to 30 minutes to complete [13, 23, 67]. DDST is a valid and reliable screening tool; although there are some reports regarding its low specificity and possible risk of overestimation [72, 73], it has a high sensitivity in detecting children with DDs [73, 74]. DDST assesses children's development through a total of 125 items generally in four domains: (1) personal-social (25 items), (2) fine motor-adaptive (29 items), (3) language (39 items), and (4) gross motor (32 items) [75, 76]. Screening by DDST generates three categories of scoring: normal, suspect, and untestable (in case the child refuses to cooperate with the examiner on items that 95% of the age-matched children successfully completed) [75].

6 Novel Biomarkers Discoveries for Screening

Along with various valid and reliable screening tools, biomarkers may be helpful in the better detection of NDDs or DDs in the future [77]. There are some ongoing studies to detect novel biomarkers that can support or suggest NDDs or DDs. Previous studies reported different metabolic patterns of intermediary metabolites, dysregulated levels of sterols, fatty acids, and phospholipids, as well as an increased level of serotonin in patients with autism spectrum disorder compared to controls [78, 79]. In a study by Kelly et al., they detected 13 and 11 plasma and stool metabolites, respectively, which were significantly different between children who were categorized to have typical development compared to those categorized as "requiring further monitoring/evaluation" based on the ASQ results. In this comprehensive metabolomics study, both the blood and stool metabolite levels showed that tryptophan and tyrosine metabolism pathways, particularly the high levels of N-formylanthranilic acid, play a crucial role in improving the communication score [80]. In another study, it was suggested that metallomics can contribute to the biomarker discoveries in attention deficit hyperactivity disorder [81]. In addition, electrophysiologic and imaging findings, eye-tracking records, actigraphy surveys, multimodal respiratory and autonomic function sensors, and individualized analysis of neural cells derived from inducible pluripotent stem cells have been suggested to be helpful in detecting children with neurodevelopmental disabilities. Integrating multiple methods rather than screening strictly with a single classical method (i.e., questionnaires and checklists) may help in initiating a more accurate diagnosis of NDDs or DDs [82].

7 Integrating Early Screening in Telemedicine

Considering the growing incidence of NDDs and DDs in the last two decades, along with discrepancies in access to primary care services, especially in low-income families, and also the COVID-19 consequences, telemedicine can promote healthcare delivery in target populations [83–86]. The limited number of pediatricians, child neurologists, or psychiatrists who are trained in early screening of NDDs and DDs, time constraints of healthcare providers, the COVID-19 outbreak and its consequences, as well as geographic isolation and resource limitations in some communities, are some other obstacles in the early detection of NDDs or DDs in early childhood, which can be solved by conducting telemedicine programs [87–93]. Previous studies showed satisfactory results regarding the use of telemedicine for early diagnosis and regular follow-up of patients with NDDs or DDs [83, 94–98]. Hence, telemedicine is an ideal and reliable service delivery platform that can be beneficial, particularly for countries with diverse levels of accessibility to healthcare systems as well as families with high-risk children.

8 Role of Parents in Early Screening

Screening by pediatricians and family physicians is time-consuming and costly [99] and monitoring strategies such as checklists and clinical observation are not sensitive enough [100]. Clinical observation without utilizing screening tools has been suggested to miss about half of the infants eligible for early intervention [101]. On the other hand, as children's development is constantly evolving, regular frequent screening is required to identify DDs and NDDs [102]. Obtaining details from parents using valid and reliable tools can provide valuable information for physicians. Parents are present during clinical visits, witness their child's abilities and behaviors that may not be visible during the physician's visit, and are not paid by the healthcare system. Moreover, most of the parent-completed tools are brief, require a brief time, and are widely available [103, 104]. The most frequently used parent-completed questionnaires that are available worldwide are the Ages and Stages Questionnaire (ASQ), Child Development Inventory (CDI), and Parents' Evaluation of Developmental Status (PEDS) [105].

Although parents play a key role in the early screening of NDDs or DDs, if a parent is frequently oblivious to an infant's relationship, fails to detect the infant's hints, is easily furious by the infant, or ignores the infant, this may be an indication of attachment difficulties. Furthermore, parents suffering from depression or substance use disorder may have difficulties fostering their child's growth and development and may benefit from assistance in providing developmental stimulation and opportunities for play and learning. Additional support services may be necessary for parents with low educational achievement or limited community resources [16].

9 Barriers to Early Screening

While a standard screening tool can identify most DDs and NDDs, it is estimated that only 23% of primary care providers utilize these tools [106]. Some of the obstacles to screening include limitations in time, financial issues, requiring human resources, lack of agreement on the most appropriate instruments, and lack of physicians' self-confidence due to inadequate training and experience [99, 105–107]. In a previous study, 82% of primary care physicians reported time constraints as the most important hindrance [23]. Lack of information and education, social stigma, limited access to primary care services in some world regions, and misunderstandings of some of the concerns reported by parents to physicians are some of the other barriers [108].

10 Conclusion

The earlier DDs or NDDs are detected, the more likely the children are to achieve their potential development. This is not possible unless parents and primary healthcare professionals are aware of the importance of early screening in children. Although, recently, the standard questionnaires are the most common methods for screening DDs or NDDs in children, recent studies have focused on some new biomarkers or novel methods to diagnose DDs or NDDs earlier and in a more accurate way. Future studies should focus on biomarker discovery and developing some methods for the early detection of DDs or NDDs in children.

Conflict of Interest None

Figures All figures were designed by the authors and are original. In case, we extracted data from a specific reference, we cited the related reference(s) in figure legends or table captions.

Funding None

References

1. Lamsal R, Dutton DJ, Zwicker JD. Using the ages and stages questionnaire in the general population as a measure for identifying children not at risk of a neurodevelopmental disorder. BMC Pediatr. 2018;18(1):122.

2. Singh A, Yeh CJ, Boone BS. Ages and stages questionnaire: a global screening scale. Bol Med Hosp Infant Mex. 2017;74(1):5–12.
3. Block RW, Dreyer BP, Cohen AR, Stapleton FB, Furth SL, Bucciarelli RL. An agenda for children for the 113th congress: recommendations from the pediatric academic societies. Pediatrics. 2013;131(1):109–19.
4. Bruder MB. Early childhood intervention: a promise to children and families for their future. Except Child. 2010;76(3):339–55.
5. Wu Q, Jia Z. Wiring the brain by clustered protocadherin neural codes. Neurosci Bull. 2021;37(1):117–31.
6. Scattolin MAA, Resegue RM, Rosário MC. The impact of the environment on neurodevelopmental disorders in early childhood. J Pediatr. 2022;98:66–72.
7. American Psychiatric Association. Diagnostic and statistical manual of mental disorders: DSM-5. Washington, DC: American Psychiatric Association; 2013.
8. Currie J, Stabile M, Manivong P, Roos LL. Child health and young adult outcomes. J Hum Resour. 2010;45(3):517–48.
9. Reichman NE, Corman H, Noonan K. Impact of child disability on the family. Matern Child Health J. 2008;12(6):679–83.
10. Lamsal R, Zwicker JD. Economic evaluation of interventions for children with neurodevelopmental disorders: opportunities and challenges. Appl Health Econ Health Policy. 2017;15(6):763–72.
11. Francés L, Quintero J, Fernández A, Ruiz A, Caules J, Fillon G, et al. Current state of knowledge on the prevalence of neurodevelopmental disorders in childhood according to the DSM-5: a systematic review in accordance with the PRISMA criteria. Child Adolesc Psychiatry Ment Health. 2022;16(1):27.
12. Astle DE, Holmes J, Kievit R, Gathercole SE. Annual research review: the transdiagnostic revolution in neurodevelopmental disorders. J Child Psychol Psychiatry. 2022;63(4):397–417.
13. Council on Children with Disabilities, Section on Developmental Behavioral Pediatrics, Bright Futures Steering Committee, Medical Home Initiatives for Children with Special Needs Project Advisory Committee. Identifying infants and young children with developmental disorders in the medical home: an algorithm for developmental surveillance and screening. Pediatrics. 2006;118(1):405–20.
14. Serrano AM, Boavida J. Early childhood Intervention: the Portuguese pathway towards inclusion. Rev Educ Inclus. 2011;4(1):123–38.
15. Bemanalizadeh M, Khoshhali M, Goli P, Abdollahpour I, Kelishadi R. Parental occupational exposure and neurodevelopmental disorders in offspring: a systematic review and meta-analysis. Current environmental. Health Rep. 2022;9:1–17.
16. Scharf RJ, Scharf GJ, Stroustrup A. Developmental milestones. Pediatr Rev. 2016;37(1):25–38.
17. Love JM, Kisker EE, Ross C, Raikes H, Constantine J, Boller K, et al. The effectiveness of early head start for 3-year-old children and their parents: lessons for policy and programs. Dev Psychol. 2005;41(6):885–901.
18. Scarborough AA, Spiker D, Mallik S, Hebbeler KM, Bailey DB Jr, Simeonsson RJJEC. A national look at children and families entering early intervention. Except Child. 2004;70(4):469–83.
19. Manning M, Homel R, Smith C, Review YS. A meta-analysis of the effects of early developmental prevention programs in at-risk populations on non-health outcomes in adolescence. Child Youth Serv Rev. 2010;32(4):506–19.
20. Manning M, Homel R, Smith C. Economic evaluation of a community based early intervention program implemented in a disadvantaged urban area of Queensland. Econ Anal Policy. 2006;36(1–2):99–119.
21. Bemanalizadeh M, Badihian N, Khoshhali M, Badihian S, Hosseini N, Purpirali M, et al. Effect of parenting intervention through "Care for Child Development Guideline" on early child development and behaviors: a randomized controlled trial. BMC Pediatr. 2022;22(1):1–9.
22. Shevell MI, Majnemer A, Rosenbaum P, Abrahamowicz M. Profile of referrals for early childhood developmental delay to ambulatory subspecialty clinics. J Child Neurol. 2001;16(9):645–50.
23. Rydz D, Shevell MI, Majnemer A, Oskoui M. Topical review: developmental screening. J Child Neurol. 2005;20(1):4–21.
24. First LR, Palfrey JS. The infant or young child with developmental delay. N Engl J Med. 1994;330(7):478–83.
25. Wolraich ML. Disorders of development and learning. Wiley Online Library; 2003.
26. Oberklaid F, Drever K. Is my child normal?: Milestones and red flags for referral. Aust Fam Physician. 2011;40(9):666–70.
27. Lakhan R, Singh A. Early intervention tool (EIT) for children with developmental delay: a pilot study. Eur Health Psychol. 2014;16(1):18–25.
28. Bang K-S. Effects of an early nursing intervention program for infants' development and mother's child rearing in poverty. J Korean Acad Nurs. 2009;39(6):796–804.
29. Canadian Task Force on Preventive Health Care. Recommendations on screening for developmental delay. CMAJ. 2016;188(8):579–87.
30. Lipkin PH, Macias MM, Baer Chen B, Coury D, Gottschlich EA, Hyman SL, et al. Trends in pediatricians' developmental screening: 2002–2016. Pediatrics. 2020;145(4):e20190851.
31. Coo H, Ouellette-Kuntz H, Lam M, Yu C. Correlates of age at diagnosis of autism spectrum disorders in six Canadian regions. Chronic Dis Inj Can. 2012;32(2):90.
32. Rosenberg SA, Zhang D, Robinson CC. Prevalence of developmental delays and participation in early

intervention services for young children. Pediatrics. 2008;121(6):e1503–e9.

33. McDonald S, Kehler H, Bayrampour H, Fraser-Lee N, Tough S. Risk and protective factors in early child development: results from the All Our Babies (AOB) pregnancy cohort. Res Dev Disabil. 2016;58:20–30.
34. Choo YY, Yeleswarapu SP, How CH, Agarwal P. Developmental assessment: practice tips for primary care physicians. Singapore Med J. 2019;60(2):57.
35. Ahmadipour S, Mohammadzadeh M, Mohsenzadeh A, Birjandi M, Almasian M. Screening for developmental disorders in 4 to 60 months old children in Iran (2015–2016). J Pediatr Neurol. 2019;17(01):008–12.
36. Valla L, Wentzel-Larsen T, Hofoss D, Slinning K. Prevalence of suspected developmental delays in early infancy: results from a regional population-based longitudinal study. BMC Pediatr. 2015;15(1):1–8.
37. Samms-Vaughan ME. The status of early identification and early intervention in autism spectrum disorders in lower- and middle-income countries. Int J Speech Lang Pathol. 2014;16(1):30–5.
38. Bornstein MH, Hendricks C. Screening for developmental disabilities in developing countries. Soc Sci Med. 1982;2013(97):307–15.
39. Squires J, Potter L, Bricker D. The ASQ user's guide for the ages & stages questionnaires: a parent-completed, child-monitoring system. Baltimore, MD; Paul H. Brookes Publishing; 1995.
40. Squires J, Bricker D. Ages & Stages Questionnaires, 3rd edn (ASQ-3). Baltimore, MD; Paul H. Brookes Publishing; 2009.
41. Vameghi R, Sajedi F, Mojembari AK, Habiollahi A, Lornezhad HR, Delavar B. Cross-cultural adaptation, validation and standardization of Ages and Stages Questionnaire (ASQ) in Iranian children. Iran J Public Health. 2013;42(5):522.
42. Sarmiento Campos JA, Squires J, Ponte J. Universal developmental screening: preliminary studies in Galicia, Spain. Early Child Dev Care. 2011;181(4):475–85.
43. Heo KH, Squires J, Yovanoff P. Cross-cultural adaptation of a pre-school screening instrument: comparison of Korean and US populations. J Intellect Disabil Res. 2008;52(3):195–206.
44. Saihong P. Use of screening instrument in Northeast Thai early childcare settings. Procedia Soc Behav Sci. 2010;7:97–105.
45. Agarwal PK, Shi L, Daniel LM, Yang PH, Khoo PC, Quek BH, et al. Prospective evaluation of the ages and stages questionnaire 3rd edition in very-low-birthweight infants. Dev Med Child Neurol. 2017;59(5):484–9.
46. Kerstjens JM, Bos AF, ten Vergert EM, de Meer G, Butcher PR, Reijneveld SA. Support for the global feasibility of the Ages And Stages Questionnaire as developmental screener. Early Hum Dev. 2009;85(7):443–7.
47. Flamant C, Branger B, Nguyen The Tich S, de La Rochebrochard E, Savagner C, Berlie I, et al. Parent-completed developmental screening in premature children: a valid tool for follow-up programs. PLoS One. 2011;6(5):e20004.
48. Translations of ASQ 2022. Available from: https://agesandstages.com/products-pricing/languages/
49. Alvik A, Grøholt B. Examination of the cut-off scores determined by the Ages and Stages Questionnaire in a population-based sample of 6 month-old Norwegian infants. BMC Pediatr. 2011;11(1):1–7.
50. Armijo I, Schonhaut L, Cordero M. Validation of the Chilean version of the Ages and Stages questionnaire (ASQ-CL) in community health settings. Early Hum Dev. 2015;91(12):671–6.
51. D'Aprano A, Silburn S, Johnston V, Robinson G, Oberklaid F, Squires J. Adaptation of the ages and stages questionnaire for remote aboriginal Australia. Qual Health Res. 2016;26(5):613–25.
52. Filgueiras A, Pires P, Maissonette S, Landeira-Fernandez J. Psychometric properties of the Brazilian-adapted version of the Ages and Stages Questionnaire in public child daycare centers. Early Hum Dev. 2013;89(8):561–76.
53. Hsiao C, Richter L, Makusha T, Matafwali B, Van Heerden A, Mabaso M, et al. Use of the ages and stages questionnaire adapted for South Africa and Zambia. Child Care Health Dev. 2017;43(1):59–66.
54. Juneja M, Mohanty M, Jain R, Ramji S. Ages and Stages Questionnaire as a screening tool for developmental delay in Indian children. Indian Pediatr. 2012;49(6):457–61.
55. Østergaard KK, Lando AV, Hansen BM, Greisen G. A Danish reference chart for assessment of psychomotor development based on the Ages & Stages Questionnaire. Dan Med J. 2012;277(14):215–96.
56. Otalvaro AR, Grañana N, Gaeto N, Torres MÁ, Zamblera MN, Vasconez MA, et al. ASQ-3: validation of the Ages and Stages Questionnaire for the detection of neurodevelopmental disorders in Argentine children. Arch Argent Pediatr. 2018;116(1):7–13.
57. Richter J, Janson H. A validation study of the Norwegian version of the Ages And Stages Questionnaires. Acta Paediatr. 2007;96(5):748–52.
58. Yao G, Bian X, Squires J, Wei M, Song W. Cutoff scores of the Ages and Stages Questionnaire-Chinese for screening infants and toddlers. Zhonghua Er Ke Za Zhi. 2010;48(11):824–8.
59. Yue A, Jiang Q, Wang B, Abbey C, Medina A, Shi Y, et al. Concurrent validity of the ages and stages questionnaire and the Bayley scales of infant development III in China. PLoS One. 2019;14(9):e0221675.
60. Yunilda E, Gunardi H, Medise BE, Oswari H. The Indonesian version of Ages and Stages Questionnaire III accuracy compared to Bayley Scales of Infant Development III. Infant Child Dev. 2022;32:e2387.

61. Hardy S, Haisley L, Manning C, Fein D. Can screening with the Ages And Stages Questionnaire detect autism? J Dev Behav Pediatr. 2015;36(7):536.
62. Sices L, Drotar D, Keilman A, Kirchner HL, Roberts D, Stancin T. Communication about child development during well-child visits: impact of parents' evaluation of developmental status screener with or without an informational video. Pediatrics. 2008;122(5):e1091–e9.
63. Velikonja T, Edbrooke-Childs J, Calderon A, Sleed M, Brown A, Deighton J. The psychometric properties of the Ages & Stages Questionnaires for ages 2–2.5: a systematic review. Child Care Health Dev. 2017;43(1):1–17.
64. Squires J, Bricker D, Twombly E. Ages & stages questionnaires: social-emotional. Baltimore: Paul H. Brookes Publishing; 2002.
65. Pooch A, Natale R, Hidalgo T. Ages and stages questionnaire: social–emotional as a teacher-report measure. J Early Interv. 2019;41(1):3–12.
66. Castellanos-Páez V, González-Gamboa D. Sensitivity and specificity of "Ages And Stages Questionnaires: social-emotional" (ASQ: SE-2) in a Colombian population. Tesis Psicológica. 2022;17(2):1–21.
67. Hamilton S. Screening for developmental delay: reliable, easy-to-use tools. J Fam Pract. 2006;55(5):415–22.
68. Schonwald A, Huntington N, Chan E, Risko W, Bridgemohan C. Routine developmental screening implemented in urban primary care settings: more evidence of feasibility and effectiveness. Pediatrics. 2009;123(2):660–8.
69. Anderson PJ, Burnett A. Assessing developmental delay in early childhood—concerns with the Bayley-III scales. Clin Neuropsychol. 2017;31(2):371–81.
70. Bode MM, D'Eugenio DB, Mettelman BB, Gross SJ. Predictive validity of the Bayley, third edition at 2 years for intelligence quotient at 4 years in preterm infants. J Dev Behav Pediatr. 2014;35(9): 570–5.
71. Balasundaram P, Avulakunta ID. Bayley scales of infant and toddler development. Treasure Island, FL; StatPearls [Internet]: StatPearls Publishing; 2021.
72. Ahsan S, Murphy G, Kealy S, Sharif F. Current developmental surveillance: is it time for change? Ir Med J. 2008;101(4):110–2.
73. Screening infants and young children for developmental disabilities. American Academy of Pediatrics Committee on children with disabilities. Pediatrics. 1994;93(5):863–5.
74. Glascoe FP. Are overreferrals on developmental screening tests really a problem? Arch Pediatr Adolesc Med. 2001;155(1):54–9.
75. Frankenburg WK, Dodds J, Archer P, Shapiro H, Bresnick B. The Denver II: a major revision and restandardization of the Denver developmental screening test. Pediatrics. 1992;89(1):91–7.
76. Kliegmann RM, Stanton B, St. Geme JW. Nelson textbook of pediatrics: first south Asia edition. Elsevier India; 2015.
77. Ruggeri B, Sarkans U, Schumann G, Persico AM. Biomarkers in autism spectrum disorder: the old and the new. Psychopharmacology. 2014;231(6):1201–16.
78. West PR, Amaral DG, Bais P, Smith AM, Egnash LA, Ross ME, et al. Metabolomics as a tool for discovery of biomarkers of autism spectrum disorder in the blood plasma of children. PLoS One. 2014;9(11):e112445.
79. Schain RJ, Freedman DX. Studies on 5-hydroxyindole metabolism in autistic and other mentally retarded children. J Pediatr. 1961;58(3):315–20.
80. Kelly RS, Boulin A, Laranjo N, Lee-Sarwar K, Chu SH, Yadama AP, et al. Metabolomics and communication skills development in children; evidence from the ages and stages questionnaire. Meta. 2019;9(3):42.
81. Scassellati C, Bonvicini C, Benussi L, Ghidoni R, Squitti R. Neurodevelopmental disorders: Metallomics studies for the identification of potential biomarkers associated to diagnosis and treatment. J Trace Elem Med Biol. 2020;60:126499.
82. Sahin M, Jones SR, Sweeney JA, Berry-Kravis E, Connors BW, Ewen JB, et al. Discovering translational biomarkers in neurodevelopmental disorders. Nat Rev Drug Discov. 2019;18(4):235–6.
83. Ciccia AH, Roizen N, Garvey M, Bielefeld R, Short EJ. Identification of neurodevelopmental disabilities in underserved children using telehealth (INvesT): clinical trial study design. Contemp Clin Trials. 2015;45:226–32.
84. Houtrow AJ, Larson K, Olson LM, Newacheck PW, Halfon N. Changing trends of childhood disability, 2001–2011. Pediatrics. 2014;134(3):530–8.
85. Kuehn BM. Data on autism prevalence, trajectories illuminate socioeconomic disparities. JAMA. 2012;307(20):2137–8.
86. Martel MM. Individual differences in attention deficit hyperactivity disorder symptoms and associated executive dysfunction and traits: sex, ethnicity, and family income. Am J Orthopsychiatry. 2013;83(2–3):165.
87. Bemanalizadeh M, Yazdi M, Yaghini O, Kelishadi R. A meta-analysis on the effect of telemedicine on the management of attention deficit and hyperactivity disorder in children and adolescents. J Telemed Telecare. 2021;30:31. https://doi.org/10.1177/1357633X211045186.
88. Heidari-Beni M, Bemanalizadeh M, Heshmat R, Qorbani M, Kelishadi R. Changes in lifestyle behaviors of children and adolescents during the COVID-19 pandemic and the impact on the development of non-communicable diseases: a narrative review. Med J Islam Repub Iran. 1936;2022(22):165.

89. Yazdi M, Bemanalizadeh M, Mohebpour F, Goli P, Daniali SS, Kelishadi R. Latent class analysis of knowledge, attitude, and practice of a population-based sample of Iranian pregnant women toward COVID-19. Adv Biomed Res. 2022;11:11.
90. Mirmohammadkhani M, Bemanalizadeh M, Yazdi M, Goli P, Mohebpour F, Saffarieh E, et al. Knowledge, attitudes, and practices toward COVID-19 among Persian birth cohort participants. J Educ Health Promot. 2021;10:10.
91. Stein RE, Horwitz SM, Storfer-Isser A, Heneghan AM, Hoagwood KE, Kelleher KJ, et al. Attention-deficit/hyperactivity disorder: how much responsibility are pediatricians taking? Pediatrics. 2009;123(1):248–55.
92. Jellinek MS. Commentary: caring for the psychosocial needs of children: from advocacy to structural change. Ambul Pediatr. 2008;8(1):8–10.
93. Tanner JL, Stein MT, Olson LM, Frintner MP, Radecki L. Reflections on well-child care practice: a national study of pediatric clinicians. Pediatrics. 2009;124(3):849–57.
94. Hill A, Theodoros D. Research into telehealth applications in speech-language pathology. J Telemed Telecare. 2002;8(4):187–96.
95. Colombo P, Buo N, Busti Ceccarelli S, Molteni M. Integrating a new online platform in primary care for early detection, referral, and intervention in autism spectrum disorder: the first Italian pivotal clinical study. Brain Sci. 2022;12(2):256.
96. Stuckey R, Domingues-Montanari S. Telemedicine is helping the parents of children with neurodevelopmental disorders living in remote and deprived areas. Taylor & Francis; 2017. p. 155–7.
97. Pearl PL, Sable C, Evans S, Knight J, Cunningham P, Lotrecchiano GR, et al. International telemedicine consultations for neurodevelopmental disabilities. Telemed e-Health. 2014;20(6):559–62.
98. Kennelly AM, McIntyre B, Wood AC, Monteiro S, Voigt RG. Patient satisfaction in neurodevelopmental pediatrics: in-person vs telemedicine. J Child Neurol. 2022;37(3):181–5.
99. Dobrez D, Sasso AL, Holl J, Shalowitz M, Leon S, Budetti P. Estimating the cost of developmental and behavioral screening of preschool children in general pediatric practice. Pediatrics. 2001;108(4):913–22.
100. Glascoe FP. Screening for developmental and behavioral problems. Ment Retard Dev Disabil Res Rev. 2005;11(3):173–9.
101. Aylward GP. Developmental screening and assessment: what are we thinking? J Dev Behav Pediatr. 2009;30(2):169–73.
102. Hyman SL, Levy SE, Myers SM, Council on Children with Disabilities, Section on Developmental and Behavioral Pediatrics. Identification, evaluation, and management of children with autism spectrum disorder. Pediatrics. 2020;145(1):e20193447.
103. Glascoe FP. Using parents' concerns to detect and address developmental and behavioral problems. J Spec Pediatr Nurs. 1999;4(1):24–35.
104. Veldhuizen S, Clinton J, Rodriguez C, Wade TJ, Cairney J. Concurrent validity of the Ages and Stages Questionnaires and Bayley Developmental Scales in a general population sample. Acad Pediatr. 2015;15(2):231–7.
105. Mackrides PS, Ryherd SJ. Screening for developmental delay. Am Fam Physician. 2011;84(5):544–9.
106. Rydz D, Srour M, Oskoui M, Marget N, Shiller M, Birnbaum R, et al. Screening for developmental delay in the setting of a community pediatric clinic: a prospective assessment of parent-report questionnaires. Pediatrics. 2006;118(4):e1178–86.
107. Oberklaid F, Efron D. Developmental delay--identification and management. Aust Fam Physician. 2005;34(9):739–42.
108. Bivarchi FA, Kehyayan V, Al-Kohji SM. Barriers to the early detection and intervention of children with autism spectrum disorders: a literature. J Nurs Educ Pract. 2021;11(11):72.

Parenting Interventions for Improving Neurodevelopment

Luciana Tornquist, Debora Tornquist, Cézane Priscila Reuter, and Roya Kelishadi

Abstract

The first years of life are considered a critical period for child development. The child's development is influenced by several factors, with parenting being a relevant factor in the care of young children. Parental interventions that encourage nurturing care have shown good results in improving children's development. Group sessions and home visits have been shown to be the most promising intervention models. Different techniques for changing behavior have shown promising results (information on posters, counseling, and providing opportunities for practice and games). These interventions have been shown to be effective in improving parenting skills and children's cognitive, motor, language, and socio-emotional development, with the most robust evidence for short-term effects. Although the evidence is less consistent, early parenting interventions with premature babies have also shown good results in the baby's cognitive and physical development and interactions.

Keywords

Child development · Parent–child relations · Early interventions · Infant · Premature

1 Introduction

Child neurodevelopment is influenced by genetic factors, health history, and family, social, and environmental factors [1]. A child's healthy development is impacted by early experiences [2], in which aspects such as nurturing care, related to good health, safety and security, adequate nutrition, responsive care, and learning opportunities, can have a positive impact [3]. In response to this wide range of early experiences, the brain evolved to adapt, which explains the rapid acquisition of language, cognitive skills, and socio-emotional skills [4].

L. Tornquist · C. P. Reuter (✉)
Graduate Program in Health Promotion, University of Santa Cruz do Sul, Santa Cruz do Sul, Brazil
e-mail: cezanereuter@unisc.br

D. Tornquist
Graduate Program in Human Movement Sciences and Rehabilitation, Federal University of Santa Maria, Santa Maria, RS, Brazil

R. Kelishadi
Child Growth and Development Research Center, Research Institute for Primordial Prevention of Non-Communicable Disease, Isfahan University of Medical Sciences, Isfahan, Iran

© The Author(s), under exclusive license to Springer Nature Singapore Pte Ltd. 2024
R. Kelishadi (ed.), *Nature, Nurture, and Neurodevelopment*,
https://doi.org/10.1007/978-981-97-7433-3_11

The first few years of life are considered a critical period for exposure to protective contextual influences that may affect long-term cognitive or psychosocial development. The prenatal period through the first 5 years appears to be a sensitive window of time for brain development and for promoting consistent, responsive, and sensitive parenting that helps provide cognitive stimulation and promote social-emotional skills [5]. On the other hand, when a child is exposed to multiple deprivations during this period, it may compromise the child's chances of reaching their full developmental potential [6].

Promoting full early childhood development requires a multisectoral approach with the creation of a framework through which development actions should be initiated and expanded, encompassing health, education, nutrition, and social protection services and programs, along with favorable policy environments, adequate funding, and intersectoral coordination. At the center of this structure is the nurturing care of young children, provided by parents, caregivers, and family members. Nurturing care should be responsive, emotionally engaging, and cognitively stimulating, involving caregivers' sensitivity to infant needs, their interactions with the child, opportunities for learning and exploration provided, and protection from harm [7].

In this context, parenting is a critical social factor for child development, as it encompasses experiences, cultural norms, and psychological factors [1]. Several factors can affect parental behaviors related to child care. These factors include education, income, and parental psychosocial health, as well as knowledge, attitudes, and beliefs associated with parenting, and caregivers' awareness of developmental processes, milestones, and children's emotional signals. In addition, the environment provided to the child, through cultural, economic, and physical characteristics, such as the availability of toys, also influences development [8].

In this way, parental interventions that seek to encourage nurturing care have been shown to be effective in improving children's development, especially among vulnerable populations [3]. These interventions include programs designed to improve parenting skills, resources, parenting knowledge and interactions, behaviors, attitudes, practices, and beliefs, usually through home visits, group sessions, or behavior change communication [4, 9].

Psychosocial stimulation interventions are increasingly being integrated into maternal and child health programs. Overall, these interventions empower caregivers on how to support their child's development through responsive and sensitive parent–child interactions [8]. Psychosocial stimulation is related to the components of "responsive care" and "early learning opportunities" included in the Nurturing Care Framework proposed by the World Health Organization. These components involve the ability of caregivers to perceive, understand, and respond to the demands of their children appropriately and to create opportunities for early learning so that the child interacts with objects, people, and environments [10].

Psychosocial stimulation can happen through physical and emotional stimuli that aim to stimulate the child's cognitive, emotional, social, and language development. Emotional stimuli happen through expressions of affection and warmth for the child, by responding to the child's needs in a timely manner, and through verbal and nonverbal communication incentives between the caregiver and the child, such as valuing or praising when the child accomplishes a task. Physical stimuli, on the other hand, are related to providing appropriate sensory experiences by providing materials or games appropriate to the child's age and by seeking to give meaning to the experiences experienced by the child [11].

2 Parental Interventions of Psychosocial Stimulation and Offspring Neurodevelopment

Consistent evidence in the literature has indicated that early childhood stimulation interventions improve several aspects related to parenting, such as the incentive to improve the home care environment, gains in caregiver–child interactions, and increased knowledge about child development. These improvements in parenting skills may serve as key mechanisms for these programs to benefit child development outcomes [12].

A review of randomized clinical trials carried out during pregnancy and in children aged 0 to 5 years showed that interventions focused on responsive care and learning opportunities had effects four to five times greater than interventions based on nutritional supplementation in cognitive, motor, and children's language skills [13]. Corroborating this, another review demonstrated that early childhood stimulation interventions had a medium effect size on cognitive and language development, while nutritional interventions had a small effect size [14].

The developmental benefits of interventions in nurturing care and parental protection are supported by the results of a meta-analysis that demonstrated the effects of these parenting programs on increasing scores on measures of child psychosocial, motor, and cognitive development. The parenting programs that proved most effective were those that applied different behavior change techniques, including information through posters and cards to illustrate positive practices, provided opportunities for the practice of games and responsive conversations with children, and offered guidance and support for change practices and problem-solving strategies [4].

A meta-analysis of randomized controlled trials based on parenting interventions for children under 2 years of age in low- and middle-income countries found small to medium overall positive effects on child developmental outcomes for cognitive, language, and motor skills. However, it found no effects under the socio-emotional domain. The interventions were based on components of responsive, stimulating, and sensitive interactions between caregiver and child. The review also demonstrated that interventions carried out through group sessions or group sessions combined with home visits were more beneficial than those carried out through isolated individual sessions, especially for cognitive and motor development [3].

These results are supported by another review that also pointed out that interventions based on group sessions and supported by home visits are the most promising program model for stimulation programs in children under 2 years of age. According to the authors, this model is less laborious, requiring fewer hours of contact with the family. In addition, it encourages and modifies child-rearing norms within the group. Stimulation-based models of interventions produced improvements in the parents' verbally responsive interactions with the child and in stimulation through the provision of materials for the child to play with. These results reflected moderate effects on cognitive and language outcomes [14].

Published reports on parenting programs in low- and middle-income countries have also found evidence of positive effects on children's cognitive and language development [15, 16]. In a report published by the United Nations Children's Fund (UNICEF), it was shown that intensive parental psychosocial stimulation programs, which took place at least once a week and for at least 1 year, were effective in changing parents' practices and children's mental, social, and emotional development outcomes. However, they were not effective for psychomotor development. Studies where programs required direct parent interaction with children were successful in improving children's information processing skills, language skills, and social and emotional well-being [15].

Similarly, a report commissioned by the UK Department for International Development (DfID) reviewed studies focusing on early childhood cognitive development in developing countries. The report found that parent-centered interventions are most effective when both the child and the parent are involved. The programs were generally implemented at home by fathers, mostly mothers, and were relatively small in scale, with an average of 62 participants. Interventions were focused on teaching parents to stimulate their children through play and focused on promoting the development of infants and children under 3 years of age. In general, short-term (1 week to 10 months) parent-focused interventions had positive effects on young children's cognitive development, while significant effects were shown in younger and older children if interventions lasted at least 2 years. Although most studies only confirmed short-term effects on young children's cognition, three interventions showed positive long-term effects [16].

The long-term effects of these interventions are still little explored in the literature. A systematic review study includes parenting interventions in low- and middle-income countries that are carried out in the first 3 years of life and that have a follow-up assessment at least 1 year after the intervention. However, the review found only three studies that conducted a mid-term follow-up evaluation (4–9 years after program completion) and only two long-term studies (greater than 10 years). The results showed a decrease in effects over time and indicated that medium- and long-term effects were largely inconclusive. The meta-analysis indicated significant effects on cognitive development post-intervention in the short term but not significant in the medium and long term. As for behavioral development, the effects were consistently null over time [17].

Interventions in early childhood with parental involvement have been shown to be effective in increasing parental sensitivity, attitudes toward sensitive discipline strategies, and child stimulation, favoring child development [18, 19]. Evidence suggests that nurturing environments that provide opportunities for children to play and communicate play a significant role in child development outcomes [3]. Britto et al. [4] point out that to promote successful and sustainable interventions, programs need to have a multisectoral focus and be supported by nurturing care. For children to reach their developmental potential, parents, caregivers, and families need to be supported in providing care and protection for the child.

3 Parental Interventions and Neurodevelopment of Preterm Infants

Premature infants may have significant short- and long-term neurodevelopmental impairments, including cognitive deficits and motor impairment [20, 21]. Children born prematurely may, due to their biological and neurological immaturity, have lower self-regulation skills, be less able to organize and modulate physiological and behavioral aspects in response to the environment, and become more passive and less communicative. In addition, infant prematurity can generate psychological effects on parents and lead them to perceive the child as a more vulnerable individual, which can make parents overprotective, leading them to interfere too much in children's demands, limiting exploration and appropriate behaviors for children, and opportunities for positive interactions [22].

Hospitalization in a Neonatal Intensive Care Unit is considered a critical period for the brain development of premature babies, and the experiences during this period can significantly influence neurodevelopment [23]. Among these experiences, social interaction with parents is one of the influential environmental factors for long-term neurobehavioral outcomes [24]. In this way, including parents in the care of premature babies is a beneficial element for supporting the proper development of babies [25].

A meta-analysis sought to determine whether early interventions aimed at teaching parenting skills and/or involving parents in the hospital care of preterm infants could improve the neurodevelopmental outcomes of preterm infants. Although some programs aimed to improve motor skills, most studies focused on cognitive development and parent–infant interactions. The applied mental development scales evaluated the child's level of cognitive development, language, and personal–social interaction. The motor scale assessed the child's fine and gross motor skills [2].

Results indicate clinically significant effects at 12 months, with significantly higher mental and physical performance scores favoring the intervention group. Improvements were also seen in mental performance scores at 24 months, but physical performance scores were not significant. At 36 months, improvements in mental performance also occurred; however, at 5 years, no improvement in neurodevelopmental outcomes is observed. In general, all studies included infants classified as low-risk. Only three of the included studies specifically assessed high-risk infants, and data were available only at 6 and 12 months, but similar effects were demonstrated [2].

The results of this meta-analysis demonstrate a positive effect of a variety of strategies aimed at promoting child development in preterm infants. There was great heterogeneity between the studies, which involved different types of interventions. Interventions were broadly classified into developmental care of a specific type known as NIDCAP, kangaroo care, and other developmental interventions, which include an in-hospital and/or post-discharge education-support-stimulation package, each with a specific component of parental involvement [2]. The Neonatal Individualized Developmental Care Assessment (NIDCAP) program is an individualized approach that is based on the observation of preterm infants' behaviors to guide care activities provided by professionals and infant families [26].

Another meta-analysis evaluated the effectiveness of interventions in the early neurodevelopment of preterm infants during their stay in the Neonatal Intensive Care Unit and after 2 weeks of corrected age at term. However, in interventions that involved parental participation, there was no significant improvement in the neurobehavioral development of preterm infants compared to standard care in preterm infants [23].

Already, a meta-analysis that verified the effects of early interventions focusing on the family on the neurodevelopment of children born preterm during the first 3 years of life found positive effects. The intervention contributed to the development of cognitive and motor domains of preterm infants. Improved cognitive development was seen at 12, 24, and 36 months. In the motor domain, the effect was observed only at 12 months [27]. A pilot study that sought to evaluate an intervention centered on parental involvement in the care of premature infants in a Chinese neonatal intensive care unit also showed a significant improvement in neurological outcomes 1 day before discharge [28].

Similarly, a randomized controlled clinical trial investigated the effectiveness of a maternal participation program on the neurobehavioral development of preterm infants admitted to a neonatal intensive care unit. The results showed better neurobehavioral development at 4 and 28 days of life among babies in the intervention group compared to the control group [25].

A meta-analysis on the impact of parenting style on cognitive and behavioral outcomes of preterm infants demonstrated that parental responsiveness and parental demand were associated with improved cognition in children. Responsiveness is related to parental sensitivity, how parents respond to the child's cues, and how well parents read the child's verbal and nonverbal cues. The requirement is characterized by the level of expectations of the parents in relation to the child, whether or not these expectations are proportional to the development, neither too high nor too low for the age and stage of the child [1].

Although the evidence is less consistent in children born prematurely than in the general pediatric population, early interventions have shown promise in improving interaction and promoting cognitive and physical development in vulnerable infants. It is necessary to deepen the knowledge about the mechanisms and specificities that impact the development of biologically vulnerable babies.

4 Conclusion

Interventions focused on improving parental skills in child-rearing care have proven to be important strategies for promoting child development. Studies have shown encouraging results, with the improvement of parental skills acting as a mechanism that benefits the development of the offspring, resulting in positive outcomes in the cognitive, motor, language, and socio-emotional domains. Models of group interventions or through home visits were more promising than individualized interventions, and different successful behavior change techniques were reported by the studies (optimizing practices, counseling, posters with good practices).

The evidence is more robust for short-term effects, and it is necessary to further investigate the effects of these interventions over longer periods. There is a need to invest in longitudinal design studies to determine whether there are any sustained benefits for children after programs end or whether reinforcement interventions are needed at a later stage. The need also emerges to establish mechanisms to control the applicability, resolution, and evaluation of existing programs. It is necessary to determine the possibilities for expanding and improving methods for future interventions so that they can develop a greater potential for changing parental behavior and sustaining the benefits of the program in the long term.

So that all children have the opportunity to develop their development potential in an equitable way, it is necessary to deepen the assessment, monitoring, and appropriate interventions, considering the local social, environmental, and economic context. Thus, considering sustainable and low-cost options for evaluating and developing interventions can reduce barriers in low- and middle-income countries.

In children born prematurely, interventions have also been shown to be effective in the cognitive and physical domains and in improving child interaction, although the evidence is less consistent. More research is needed to determine the most effective interventions for premature infants and for biologically vulnerable infants.

References

1. Neel MLM, Stark AR, Maitre NL. Parenting style impacts cognitive and behavioral outcomes of former preterm infants: a systematic review. Child Care Health Dev. 2018;44:507. https://doi.org/10.1111/cch.12561.
2. Vanderveen JA, Bassler D, Robertson CMT, Kirpalani H. Early interventions involving parents to improve neurodevelopmental outcomes of premature infants: a meta-analysis. J Perinatol. 2009;29:343–51. https://doi.org/10.1038/JP.2008.229.
3. Zhang L, Ssewanyana D, Martin MC, et al. Supporting child development through parenting interventions in low- to middle-income countries: an updated systematic review. Front Public Health. 2021;9:950. https://doi.org/10.3389/fpubh.2021.671988/bibtex.
4. Britto PR, Lye SJ, Proulx K, et al. Nurturing care: promoting early childhood development. Lancet. 2017;389:91–102. https://doi.org/10.1016/S0140-6736(16)31390-3.
5. Wachs TD, Georgieff M, Cusick S, Mcewen BS. Issues in the timing of integrated early interventions: contributions from nutrition, neuroscience, and psychological research. Ann N Y Acad Sci. 2014;1308:89–106. https://doi.org/10.1111/NYAS.12314.
6. Schiariti V, Simeonsson RJ, Hall K. Promoting developmental potential in early childhood: a global framework for health and education. Int J Environ Res Public Health. 2021;18(4):2007. https://doi.org/10.3390/ijerph18042007.
7. Richter LM, Daelmans B, Lombardi J, et al. Investing in the foundation of sustainable development: pathways to scale up for early childhood development. Lancet. 2017;389:103–18. https://doi.org/10.1016/S0140-6736(16)31698-1.
8. Bentley ME, Johnson SL, Wasser H, et al. Formative research methods for designing culturally appropriate, integrated child nutrition and development interventions: an overview. Ann N Y Acad Sci. 2014;1308:54–67. https://doi.org/10.1111/nyas.12290.
9. Engle PL, Black MM, Behrman JR, et al. Strategies to avoid the loss of developmental potential in more than 200 million children in the developing world. Lancet. 2007;369:229–42. https://doi.org/10.1016/S0140-6736(07)60112-3.
10. World Health Organization. Nurturing care for early childhood development: a framework for helping children survive and thrive to transform health and human potential. Geneva: World Health Organization; 2018.
11. World Health Organization. Mental health and psychosocial well-being among children in severe food shortage situations. Geneva: World Health Organization, Mental Health: Evidence and Research Department of Mental Health and Substance Abuse; 2006.
12. Jeong J, Pitchik HO, Yousafzai AK. Stimulation interventions and parenting in low- and middle-income countries: a meta-analysis. Pediatrics. 2018;141:e20173510. https://doi.org/10.1542/peds.2017-3510.
13. Prado EL, Larson LM, Cox K, et al. Do effects of early life interventions on linear growth correspond to effects on neurobehavioural development? A systematic review and meta-analysis. Lancet Glob Health. 2019;7:e1398–413. https://doi.org/10.1016/S2214-109X(19)30361-4.
14. Aboud FE, Yousafzai AK. Global health and development in early childhood. Ann Rev Psychol. 2015;66:433–57. https://doi.org/10.1146/annurev-psych-010814-015128.
15. Britto PR, Ponguta LA, Reyes C, Karnati R. A systematic review of parenting programs for young children. New York: United Nations Children's Emergency Fund; 2015.
16. Rao N, Sun J, Wong JMS, et al. Early childhood development and cognitive development in developing countries: a rigorous literature review. London: Department for International Development; 2014.

17. Jeong J, Pitchik HO, Fink G. Short-term, medium-term and long-term effects of early parenting interventions in low- and middle-income countries: a systematic review. BMJ Glob Health. 2021;6:e004067. https://doi.org/10.1136/bmjgh-2020-004067.
18. Bakermans-Kranenburg MJ, Van IJzendoorn MH, Juffer F. Less is more: meta-analyses of sensitivity and attachment interventions in early childhood. Psychol Bull. 2003;129:195–215. https://doi.org/10.1037/0033-2909.129.2.195.
19. Van Zeijl J, Mesman J, Van Ijzendoorn MH, et al. Attachment-based intervention for enhancing sensitive discipline in mothers of 1- to 3-year-old children at risk for externalizing behavior problems a randomized controlled trial. J Consult Clin Psychol. 2006;74:994–1005. https://doi.org/10.1037/0022-006X.74.6.994.
20. Pickler RH, McGrath JM, Reyna BA, et al. A model of neurodevelopmental risk and protection for preterm infants. Adv Neonatal Care. 2013;13 Suppl 5:S11–20. https://doi.org/10.1097/ANC.0000000000000022.
21. Volpe JJ. Brain injury in premature infants: a complex amalgam of destructive and developmental disturbances. Lancet Neurol. 2009;8:110. https://doi.org/10.1016/S1474-4422(08)70294-1.
22. Toscano C, Soares I, Mesman J. Controlling parenting behaviors in parents of children born preterm: a meta-analysis. J Dev Behav Pediatr. 2020;41:230–41. https://doi.org/10.1097/DBP.0000000000000762.
23. Aita M, de Clifford Faugère G, Lavallée A, et al. Effectiveness of interventions on early neurodevelopment of preterm infants: a systematic review and meta-analysis. BMC Pediatr. 2021;21:1–17. https://doi.org/10.1186/S12887-021-02559-6/figures/6.
24. Santos J, Pearce SE, Stroustrup A. Impact of hospital-based environmental exposures on neurodevelopmental outcomes of preterm infants. Curr Opin Pediatr. 2015;27:254–60. https://doi.org/10.1097/MOP.0000000000000190.
25. Namprom N, Picheansathian W, Jintrawet U, Chotibang J. The effect of maternal participation in preterm's care and improved short-term growth and neurodevelopment outcomes. J Neonatal Nurs. 2018;24:196–202. https://doi.org/10.1016/j.jnn.2018.03.003.
26. Ohlsson A, Jacobs SE. NIDCAP: a systematic review and meta-analyses of randomized controlled trials. Pediatrics. 2013;131:e881–93. https://doi.org/10.1542/peds.2012-2121.
27. Ferreira R d C, Alves CRL, Guimarães MAP, et al. Effects of early interventions focused on the family in the development of children born preterm and/or at social risk: a meta-analysis. J Pediatr. 2020;96:20–38. https://doi.org/10.1016/j.jped.2019.05.002.
28. Zhang R, Huang RW, Gao XR, et al. Involvement of parents in the care of preterm infants: a pilot study evaluating a family-centered care intervention in a Chinese neonatal ICU. Pediatr Crit Care Med. 2018;19:741–7. https://doi.org/10.1097/PCC.0000000000001586.

GPSR Compliance

The European Union's (EU) General Product Safety Regulation (GPSR) is a set of rules that requires consumer products to be safe and our obligations to ensure this.

If you have any concerns about our products, you can contact us on ProductSafety@springernature.com

In case Publisher is established outside the EU, the EU authorized representative is:

Springer Nature Customer Service Center GmbH
Europaplatz 3
69115 Heidelberg, Germany

Batch number: 10731385

Printed by Printforce, the Netherlands